AF662768

Optical Communication

Optical Communication

M Mukunda Rao

Universities Press

Universities Press (India) Private Limited

Registered Office
3-6-747/1/A & 3-6-754/1, Himayatnagar, Hyderabad 500 029 (A.P), India
e-mail: *www.universitiespress.com*

Distributed by
Orient Blackswan Private Limited

Registered Office
3-6-752, Himayatnagar, Hyderabad 500 029 (A.P), India

Other Offices
Bangalore / Bhopal / Bhubaneshwar / Chennai /Ernakulam / Guwahati
Hyderabad / Jaipur / Kolkata /Lucknow / Mumbai / New Delhi / Patna

© Universities Press (India) Private Limited 2000
First published 2000
Reprint 2009, 2013

ISBN: 978 81 7371 090 2

Typeset by
OSDATA,
Hyderabad 500 029

Printed at
Bluebells Secure Prints
Hyderabad 500 072

Published by
Universities Press (India) Private Limited
3-6-747/1/A & 3-6-754/1, Himayatnagar,
Hyderabad 500 029 (A.P), India

This book is dedicated to the

fond memory of

Professor B. M. Sivaram (1941–1999)

Foreword

I am very happy and delighted to write the Foreword for a textbook on Optical Communication written by Prof. Mukunda Rao.

The text is divided into nine easily readable chapters starting with an Introduction to the subject and ending with future trends. The subject material, divided into well-organised chapters, deals with optical sources (transmitters), medium of transmission (optical fiber), and photodetectors (receivers). The modulation and demodulation techniques suitable for optical communication are dealt with separately. Numerical examples and problems have been included wherever required for a proper appreciation of the subject.

I am sure this book will go a long way in filling the vacuum of good textbooks for students of this interesting and fast developing subject.

Dr. Ing. B.V.A. Rao
Principal, Vellore Engineering College

Preface

Optical Communication is a relatively new field which is still in the process of emerging as a full-fledged technology. The developments in this field are indeed breath-taking with major communication industries fanning its growth through their R&D efforts. Besides, many academic institutions are also involved in the development of this fascinating field, both in terms of basic research for the academic challenges it offers, and applied research for the funding it can generate from both governmental and non-governmental agencies. However, the locomotive for its enormous growth during the last couple of decades has essentially been the initiative taken by the private telecom industry which could perceive correctly the enormous growth potential it offers for increased and improved telecom services to their customers. Thus, the need and demand for trained personnel in this area was increasingly felt. As an answer to this, Optical Communication was introduced in the curricula of Electronics and Communication Engineering faculties, to begin with at the IITs and IISc, and gradually at other universities. In view of these developments, the need for a comprehensive textbook on optical communication emerged, tailored to the needs of classroom teaching. Though there are excellent books on Optical Communication available in the market, they are not designed for classroom teaching. The teacher is required to have a standard textbook which is based on the syllabus at Indian universities, and which serves as a source for day-to-day teaching and learning by the teacher and the taught.

Since the author has the distinction of having introduced this subject at the Indian Institute of Technology, Chennai, in the 1980s, and taught this subject for more than a decade to the Electronics and Communication Engineering students of that Institute, he felt the need for preparing such a textbook which could be of use for future generations of teachers and students who are obliged to take this subject. Drawing from his teaching experience, he has attempted to create a comprehensive textbook which includes numerical problems, both solved and unsolved.

It is hoped that this book will serve the purpose for which it is intended.

Contents

Foreword vii

Preface ix

1. Introduction **1**

1.1 Historical background 1
1.2 Sources for optical communication 4
1.3 Propagation factors in optical communication 6
1.4 Guided wave propagation: fiber optics 8
1.5 Photodetectors and optical receivers 11
1.6 Fiber-optic communication systems 13
1.7 Integrated optics 14
1.8 Future trends 15

2. Optical sources: The LASER **17**

2.1 Historical background 17
2.2 Laser fundamentals 18
2.3 Einstein's A & B coefficients 19
2.4 Population inversion and gain medium 21
2.5 Optical feedback and laser oscillation 24
2.6 Optical resonator: The stability condition 27
2.7 Laser system 29
2.8 Laser characteristics 32

3. Optical sources: The semiconductor laser diode and light emitting diode **35**

3.1 Historical background 35
3.2 Radiative recombination in semiconductor materials 36
3.3 The p–n junction 37

3.4 Degenerate semiconductor diode 39
3.5 Estimation of forward bias voltage 41
3.6 Threshold current density 43
3.7 Heterostructure devices 47

4. Transmission medium: Atmospheric propagation **51**

4.1 Historical background 51
4.2 Laser beam launching into the atmosphere 51
4.3 Atmospheric attenuation 59
4.4 Atmospheric turbulance 61
4.5 Parameters of atmospheric propagation 62

5. Transmission medium: Fiber optics **64**

5.1 Historical background 64
5.2 Optical fiber attenuation 64
5.3 Principle of fiber optics 66
5.4 Group delay in step-index fibers 71
5.5 Waveguide equations 73
5.6 Wave equations for step-index fiber 75
5.7 Single-mode step-index optical fiber 79
5.8 Graded-index optical fibers 81
5.9 Summary 82

6. Optical fiber characterization and fabrication **84**

6.1 Historical background 84
6.2 Fiber-optic attenuation as a function of wavelength 84
6.3 Pulse delay and distortion in a single-mode fiber 86
6.4 Fiber optic characteristics 90
6.5 Connectors, couplers and splicing 94
6.6 Fiber characterization 99

7. Photodetectors and optical receivers **103**

7.1 Historical background 103
7.2 Principles of photodetection 103
7.3 Photoemissive detectors 105

7.4 Semiconductor photodetectors 106
7.5 Optical receivers 112
7.6 Digital optical receivers 114

8. Modulation and demodulation schemes in optical communication 118

8.1 Historical background 118
8.2 Fundamentals of modulation schemes 118
8.3 Modulation spectral analysis 119

8.3.1 Analog modulation schemes 120
8.3.2 Pulse modulation schemes 121
8.3.3 Digital modulation schemes 122

8.4 Demodulation methods 124

8.4.1 Direct detection methods 124
8.4.2 Heterodyne detection receiver 125
8.4.3 Homodyne detection optical receiver 126
8.4.4 Modulation parameter convertors 127

8.5 External optical modulators 127

8.5.1. Electro-optic modulator 128

8.6 Acousto-optical modulation 134
8.7 Modulation of semiconductor devices 136

8.7.1 Modulation circuits for light emitting diodes 137
8.7.2 Laser diode modulation circuits 139

9. Optical communication systems 141

9.1 Historical background 141
9.2 Design issues in a fiber-optic communication link 142

9.2.1 Power budget 144

9.3 Time budget 145
9.4 Optical repeater and amplifiers 147
9.5 Coherent optical systems 148

9.6 Satellite optical communication systems 150
9.7 Terrestrial atmospheric optical communication lines 152
9.8 Underwater optical communication systems 153

9.8.1 Fiber-optic trans-oceanic cable systems 153
9.8.2 Optical communication with submerged bodies 154

9.9 Current developments 157

9.9.1 Integrated optics 157
9.9.2 Soliton propagation in optical fibers 158
9.9.3 New materials and devices 159

9.10 Future trends 160

Bibliography *163*

Physical constants *165*

Index *167*

1 Introduction

1.1 Historical background

The field of optical communication, using light as the carrier of information, is not new. It is a historical fact that the Greeks used fire signals in the eighth century BC for sending alarm calls for help and announcement of certain events. Only one type of signal was used, its meaning being pre-arranged between the sender and the receiver. It is believed that during the pre-Christian era of the Graeco–Persian wars, the Persian army used smoke signals to communicate, while the Greeks largely depended upon (marathon) runners. The journey of the three wise men from the East was guided by starlight in finding the child Christ. During the Napoleonic wars in Europe, a French army engineer had designed and developed an optical communication network using light towers built on hill-tops for sending/receiving signals. The optical network was so successful that it was largely responsible for Napoleon's success in the battlefield by ensuring quick movement of his army to far-flung areas. The white settlers arriving in North America had no clue about the effective communication network used by the Red Indian tribes in the form of smoke signals. The Germans had developed a 'lichtsprecher'—light speaker—for communication for their armed deployments along the Normandy coast while defending Allied invasion. For many decades now, sailors have been successfully using light blinkers to transmit messages from ship-to-ship, ship-to-shore and shore-to-ship.

However, the most significant optical communication system development came from Alexander Graham Bell, the inventor of the telephone. The optical communication system developed by Bell closely resembles the present-day optical communication system. Besides approaching the problem in his characteristic, systematic and scientific way, Bell was involved in the project in an emotional way. He patented his invention under the name 'photophone' meaning a telephone system using photons. However, unlike his celebrated telephone system over which people fought for patent rights, the photophone attracted very little commercial interest. It is believed that Bell spent more time on the development of the photophone which brought him no returns, than on the development of the telephone, which brought him fame and fortune. Bell's identification with the development of the photophone was so total that he, with the help of his assistant Watson, spent almost two years experimenting with it on the streets of Washington D.C.. He was so engrossed in these experiments that a story goes that he had to be persuaded several times to visit his new-born child in his native town. Having visited his family and seeing his baby daughter, he wanted to name her 'Photophone'!

We will now take a close look at the photophone system which is given in Fig.1.1, with a view to examine the lacunae in it. The serious limitation in this system is that it depends on sunlight as the medium of information transmission. This implies that the system cannot be operated during nights and during bad weather conditions. Besides, sunlight is incoherent, and therefore does not exhibit stabilization of wavelength and phase. This, in turn, makes it very difficult to transmit information and retrieve it at the other end. So, the need for an artificial coherent source of optical radiation became necessary. This took almost a century to fructify, with the advent of the 'LASER' in 1960. The laser (discovered by Theodore H. Maiman) was hailed as a pivotal discovery of the century, and promised many possible applications in many fields. However, when asked by press reporters to name the most important aspect of its applications, Maiman prophetically predicted that it would soon revolutionize the field of communications. The communication engineering community was quick to grasp the significance of the laser for optical communication. Many communication engineers/scientists in the early 1960s started seriously looking at the possibility of utilising this new-found device as a coherent transmitter in an optical communication system. The enthusiasm for realizing a practical/operational optical communication system was based on hard economic factors. Figure 1.2 shows the electromagnetic spectrum and the range of frequencies/wavelengths utilised for optical communication.

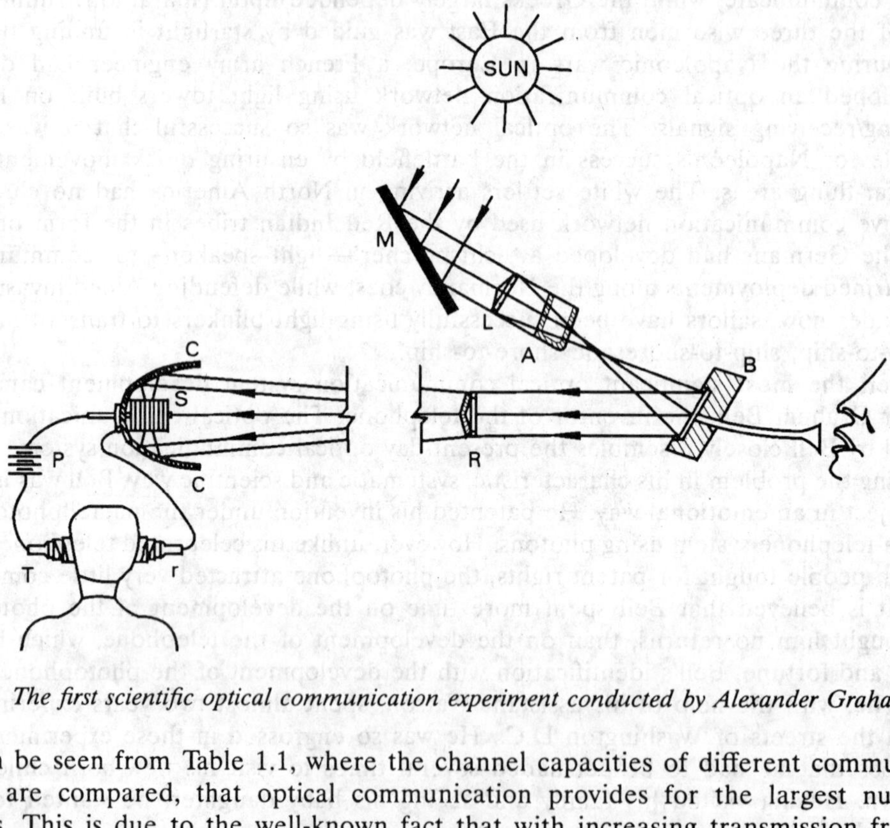

Fig. 1.1 The first scientific optical communication experiment conducted by Alexander Graham Bell.

It can be seen from Table 1.1, where the channel capacities of different communication systems are compared, that optical communication provides for the largest number of channels. This is due to the well-known fact that with increasing transmission frequency, the bandwidth of a communication system will increase. However, this is subject to the exploitation of the theoretically available bandwidth at higher frequencies. This has created serious problems in technical realization because of the peculiarities of laser construction as against the electromagnetic wave generators at lower frequencies. These details will be discussed separately in the following chapters.

Table 1.1. *A comparison of the number of telephone and TV channels that can be transmitted on different types of carriers.*

Carrier type	Frequency range	Usable B.W.	Appropriate No. of telephone channels	Appropriate No. of TV channels
LW	30 kHz–300 kHz	10%	3	-
MW	300 kHz–3 MHz	10%	25	-
SW	3 MHz–30 MHz	10%	200	-
VHF	30 MHz–300 MHz	10%	4000	-
UHF	300 MHz–3 GHz	10%	10^4	10
μ-Wave	3 GHz–1 THz	10%	10^5	100
Optical	5×6^{13} Hz–10^{15} kHz	0.1%	10^8	10^5

Assume: Telephone channel B.W. = 5 kHz; TV channel B.W. = 8 MHz

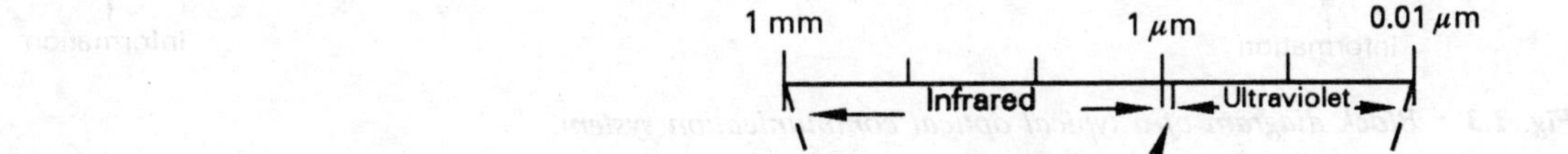

Fig. 1.2 Electromagnetic spectrum showing the range of optical communication.

1.2 Sources for optical communication

A typical optical communication system is shown in Fig. 1.3. The early laser systems like the Ruby laser or the He–Ne laser, were not amenable for direct modulation at high frequencies. Hence they required indirect or external modulation. This brought along with it cost, complexity and other complications, making the system impractical for field operations. Thus the development of a directly modulatable, compact and robust laser system became imperative for the success of optical communication.

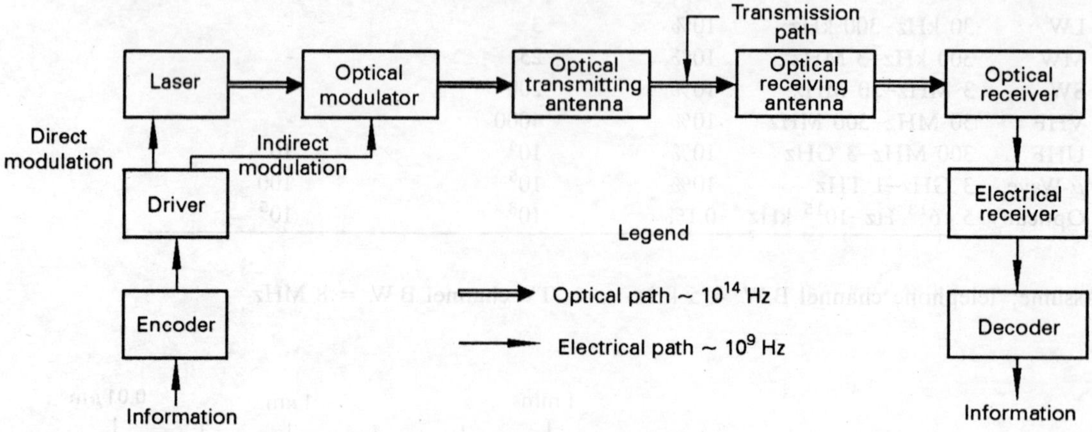

Fig. 1.3 *Block diagram of a typical optical communication system.*

The first observations of the stimulated emission of light from direct bandgap semiconductor (e.g. GaAs) *p–n* junction diodes were reported in 1962. A laser diode must satisfy the following three basic requirements

1. It must be constructed from a direct bandgap semiconductor.
2. It must have a Fabry–Perot cavity, consisting of two mirror-like surfaces defining the direction of the photon flux.
3. It must have a region which confines the radiation and the injected carriers.

The energy level diagrams of indirect bandgap materials (e.g. Si) and direct bandgap materials (e.g. GaAs) show that GaAs and other direct bandgap materials are more suitable for laser diodes because of their higher recombination probabilities. The Fabry–Perot cavity is generally created by cleaving the material along parallel crystal planes, either of which may be coated, to a final reflectivity R_1 and R_2 as shown in Fig.1.4. For lasing to take place, a sufficiently high concentration of carriers must be injected into the active region to induce population inversion. Hence the term *injection laser*.

Both carrier confinement and photon confinement take place at the junction in the dielectric waveguide d formed by the dielectric constants ϵ_2, ϵ_1, and ϵ_3 as shown in the figure. Of all the types of lasers, the semiconductor Injection Laser Diode (ILD) is exceptionally suited for optical communication because it is directly modulatable, besides being small and rugged.

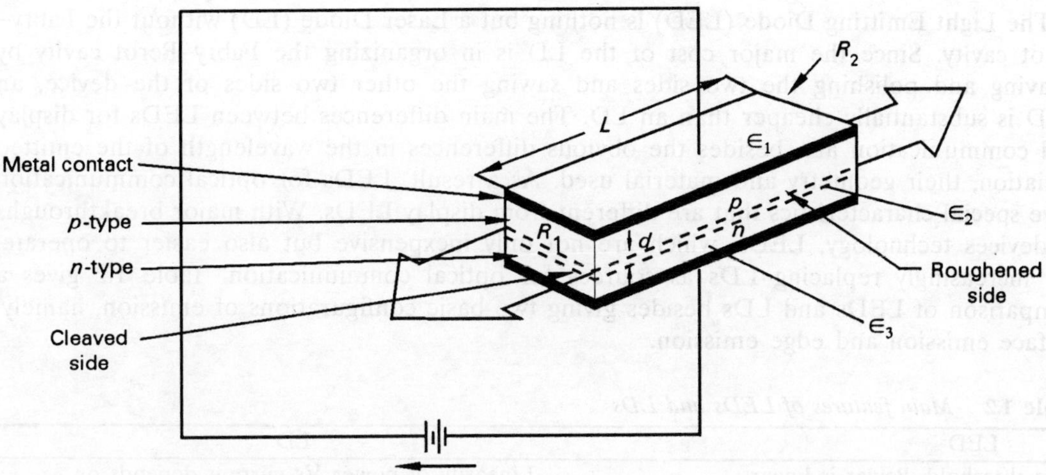

Fig. 1.4 GaAs homostructure laser.

Besides creating the resonator cavity in one direction by cleaving, the GaAs material in the other direction in the horizontal plane is deliberately sawn so that photon build-up in that direction is inhibited. Thus, the pumping of charged carriers in the vertical plane by an external bias voltage will result in laser light in one direction in the horizontal plane as shown in Fig.1.4. There is a minimum threshold current J_{th}, which has to be passed through the junction to overcome the losses as shown in Fig. 1.5. Below the threshold current, it is essentially a random spontaneous emission of photons. Laser action due to stimulated emission will result only above the threshold current due to sufficient gain in the active medium to overcome the losses, resulting in the spatial narrowing of the emitted radiation.

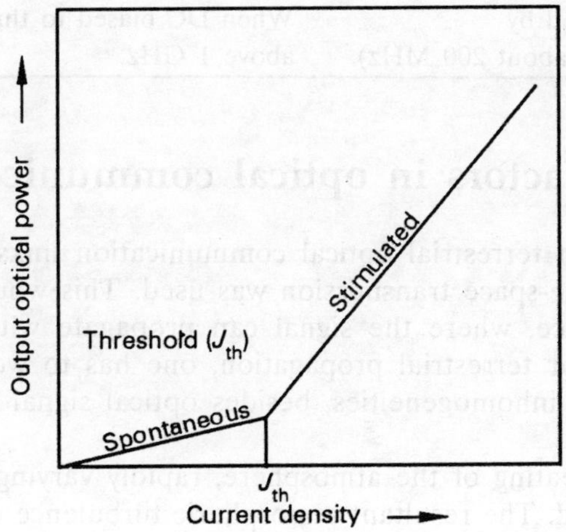

Fig. 1.5 The radiation power output as the current density in the laser increases. J_{th} is the threshold current density for laser action.

The Light Emitting Diode (LED) is nothing but a Laser Diode (LD) without the Fabry–Perot cavity. Since the major cost of the LD is in organizing the Fabry–Perot cavity by cleaving and polishing the two sides and sawing the other two sides of the device, an LED is substantially cheaper than an LD. The main differences between LEDs for display and communication are, besides the obvious differences in the wavelength of the emitted radiation, their geometry and material used. As a result, LEDs for optical communication have special characteristics that are different from display LEDs. With major breakthroughs in devices technology, LEDs, which are not only inexpensive but also easier to operate, are increasingly replacing LDs as sources for optical communication. Table 1.2 gives a comparison of LEDs and LDs besides giving two basic configurations of emission, namely, surface emission and edge emission.

Table 1.2 *Main features of LEDs and LDs*

LED	LD
No threshold. Power is linear with current.	Linearity of power Vs current depends on operating conditions; linear region above threshold only.
Less critical construction (cheaper than LD).	More critical construction for high peıfomance; increases cost.
Small temperature dependence of output.	Threshold may double between 20 and 70°C. Requires temperature stabilization or optical feedback to control output.
Emission spectrum broad (300–500 Å); increases dispersion.	Emitted linewidth is under 20 Å.
Radiance relatively low; reduces power into fiber.	High radiance—milliwatts—into low NA fibers easily obtained.
Modulation capability limited by carrier lifetime (good upto about 200 MHz).	When DC biased to threshold, can be modulated above 1 GHz.

1.3 Propagation factors in optical communication

The initial experiments on terrestrial optical communication links were modelled on the microwave links where free-space transmission was used. This would be simple and probably feasible in outer space, where the signal can propagate without hindrance, due to high vacuum. However, for terrestrial propagation, one has to worry about scattering by dust particles and density inhomogeneities, besides optical signal attenuation by selective atmospheric absorption.

Because of the solar heating of the atmosphere, rapidly varying density fluctuations on a small scale are generated. The resultant atmospheric turbulence can cause several effects which are presented in Table 1.3. Besides, the selective absorption by atmospheric gases will result in transmissivity bands as shown in Fig. 1.6. This will imply that only certain lasers can be used. Also, weather formations can cause serious absorption; for example, fog can easily cause a 40–60 dB loss. In short, free-space transmission is impractical, except for a few specialized applications. Perhaps the earliest recorded evidence of guided wave propagation is the apparatus patented by William Wheeler in 1881 for distributing lighting through a multistorey building from a central source. The first attempt at guided wave propagation for optical communication was to use a series of converging lenses, each of focal length f (several tens of metres) and placed $4f$ apart. A source, $2f$ in front of one lens, is imaged to a point $2f$ behind it, which is exactly $2f$ in front of the next lens, and so on. The system was studied in the 1960s, but it was found that the propagation was so affected by the atmosphere that even for a small distance, say a few hundred metres, effective propagation was possible only at nights, when the air was still.

Table 1.3 *Consequences of atmospheric turbulence on laser communication.*

Beam steering	Angular deviation of the beam from the line of sight path, causing the beam to miss the receiver.
Image dancing	Variations in the beam arrival angle, causing the focus point to move in the image plane.
Beam spreading	Small angle scattering, increasing the beam divergence and causing a decrease in spatial power density at the receiver.
Beam scintillation	Small-scale destructive interference within the beam cross section, causing variations in the spatial power density at the receiver.
Spatial coherence degradation	Losses in phase coherence across the phase fronts, degrading the photomixing performance.
Polarization fluctuations	Variations in the polarization state.

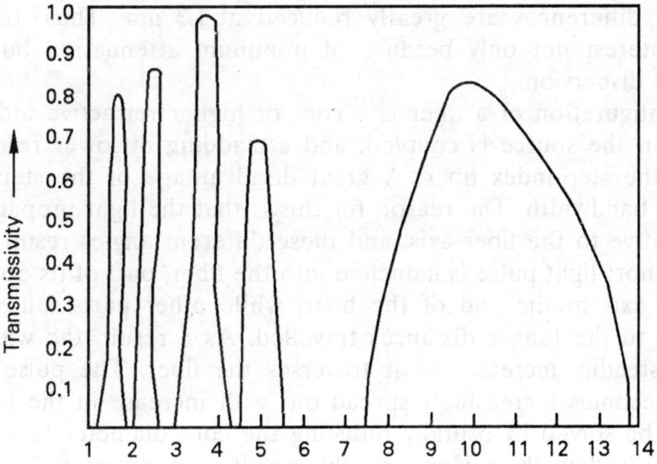

Fig. 1.6 *Principal transmissivity bands of the clear atmosphere for radiation of wavelength 1–15 μm.*

1.4 Guided wave propagation: fiber optics

If light emerging from a glass with higher refractive index n_1 is incident on a glass with lower refractive index n_2, it is essentially confined in the region of higher refractive index as long as the angle of incidence at the boundary is greater than the critical angle for total internal reflection. Using this concept, dielectric waveguides using glass fibers were experimented with in the early 1960s as the transmission medium for optical communication. However, it was found that the material absorption in glass is heavy, leading to optical attenuation factors of well over 100 dB/km. Besides, in this configuration, even very small bends would cause excessive radiation loss. Thus, the optical fiber structure as a medium of transmission is more attractive provided the material loss could be sufficiently reduced. It has been identified that the large losses in fibers were, in fact, due to absorption by impurity ions (Fe, Co, Ni, Cr, etc.) present in glass. If all these are removed, it can be theoretically shown that the material loss can be as low as 1 dB/km at 1 μm wavelength. This remaining intrinsic loss is due to Rayleigh scattering by inhomogeneities frozen into the supercooled liquid state.

The important characteristics of an optical fiber are: loss (attenuation), bandwidth (dispersion), ease of handling, and durability. The Rayleigh scattering losses in the fiber decrease with increase in wavelength, following the λ^{-4} law. However, the absorption losses are complex functions of the wavelength as shown in Fig. 1.7. Thus the choice of wavelength for optimum optical transmission becomes important, leading to the selection of optical windows at 0.85 μm, 1.3 μm and 1.55 μm. (Optical windows are dealt with in greater detail later in the book.) Another factor which has to be borne in mind while selecting the operational wavelength is the material dispersion. Material dispersion arises because of the fact that the refractive index n of glass is not constant but decreases gradually with increasing optical wavelength λ. As a result, the velocity of light in glass becomes wavelength-dependent. Since the light from optical transmitters has a certain spectral width, delay differences occur even within a single mode. At 0.85 μm, the delay differences that occur are approximately 0.1 ns per nanometre of spectral width per kilometre of fiber length. The delay differences are greatly reduced at 1.3 μm. Thus, the optical window at 1.3 μm is of interest not only because of minimum attenuation but also because of minimum material dispersion.

The normal configuration of a fiber is a core of higher refractive index into which the optical energy from the source is coupled, and a cladding of lower refractive index. This is referred to as the step-index fiber. A great disadvantage of the step-index fiber is its small transmission bandwidth. The reason for this is that the light propagates in its core at various angles relative to the fiber axis, and these different angles result in varying optical path lengths. If a short light pulse is launched into the fiber, part of its optical power travels directly along the axis to the end of the fiber, while other parts follow zig-zag courses, arriving later, due to the longer distances travelled. As a result, the width of the initially short light pulse steadily increases as it traverses the fiber. The pulse appearing at the end of the fiber becomes increasingly spread out with increase in the length of the fiber. This problem can be solved by brutally reducing the core diameter to such an extent that only one mode can be launched. However, this results in a dramatic loss of optical energy that could possibly be coupled into the fiber from the source. This disadvantage is largely overcome if the refractive index in the fiber cores is made to decrease parabolically with the radial distance, from a maximum at the fiber axis to a minimum value in the cladding. In a graded-index fiber of this type, the light, instead of zig-zagging, follows an undulating course with a period of a few millimetres. The greater length of the most devious trajectories is compensated by the light outside the fiber axis travelling faster due to the lower refractive index, so that all light components take virtually the same time to reach the end of the fiber. Graded-index fibers can now be fabricated, whose refractive index profile comes very close to the ideal.

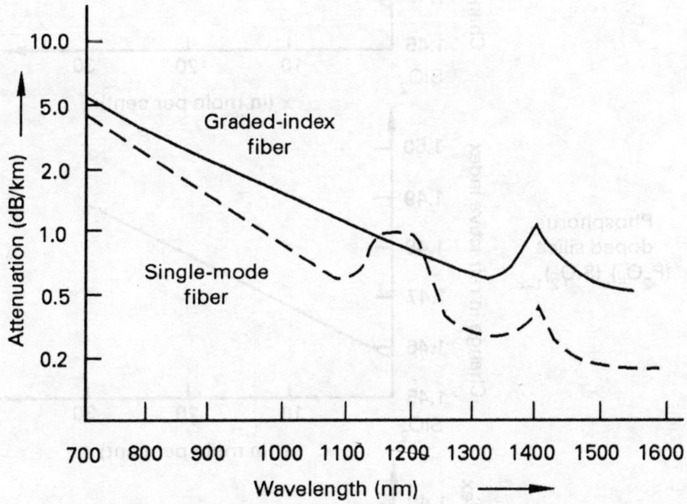

Fig. 1.7 Attenuation spectrum for multimode and single-mode fiber.

Optical waveguides for communication consist mainly of silica glass (silicon dioxide, SiO_2) with the light-guiding fiber core doped by addition of germanium and phosphorus to increase the refractive index and possibly boron or fluorine to reduce the refractive index. The change in refractive index of the silica glass as a function of the dopant concentration is shown in Fig.1.8. The outstanding suitability of SiO_2 for preparing high-purity, ultra-transparent glasses is due not only to the low intrinsic absorption (which can also be obtained from common multicomponent silicate glasses) but more specifically to the fact that silica glass—doped or undoped—can be obtained by deposition from a pure and homogeneously mixed gas phase. There are three main techniques for fabricating fibers. Most of the high-performance fibers are presently made using vapour phase reactions. The cross-sectional illustration of different fiber types and their associated index profiles are given in Fig.1.9.

Once the fiber is drawn, it is characterized. The parameters which should be measured include attenuation, dispersion, bandwidth, refractive index profile and geometry. Several strands of the fiber are now suitably arranged in arrays for mass termination techniques before they are strengthened and jacketed for underground deployment. A comparison of the attenuation of the optical fiber as a function of modulation bandwidth with that of standard electrical cables shows that at higher modulation bandwidth, the fiber-optic cables are much more advantageous. Thus, fiber optics offers several advantages for guided wave propagation of optical frequencies due to more compact size, lower cost, EMI free, higher reliability and upgradability.

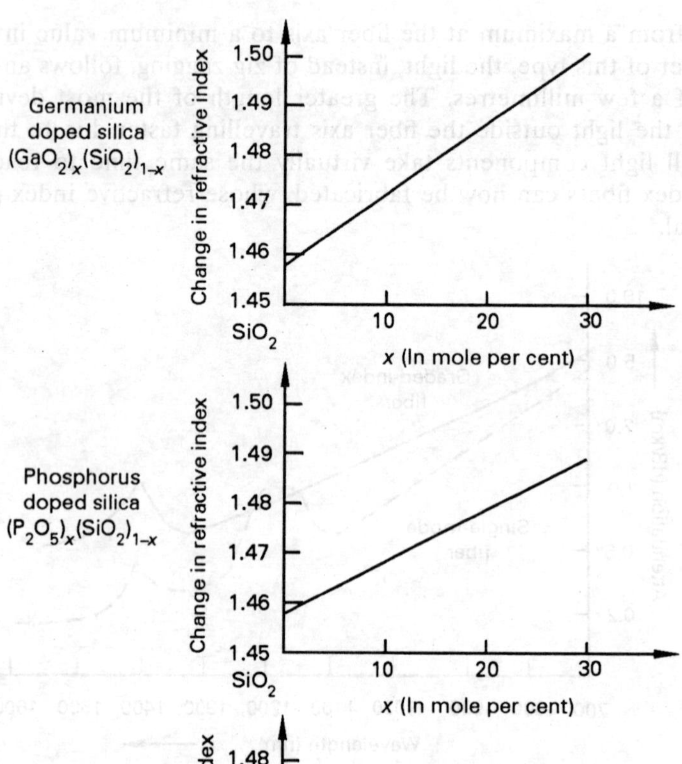

Fig. 1.8 *Refractive index for doped silica glass.*

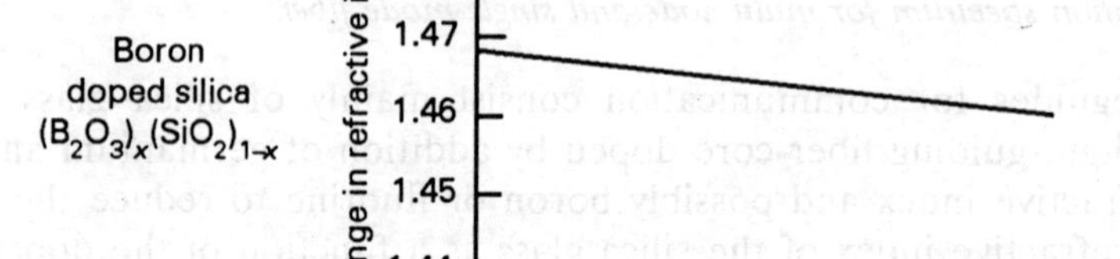

Fig. 1.9 *Cross-sectional illustration of different fiber types and their associated index profiles.*

1.5 Photodetectors and optical receivers

Photodetectors are at the heart of the optical receivers that retrieve electrical signals from optical carriers. Among the various photodetectors, the photodiode, wherein a reverse biased *p–n* junction converts the photon into a hole–electron pair, proved to be most popular. Silicon photodiodes are particularly suitable for the optical fiber transmission window in the wavelength region from 800 to 900 nm. Their spectral responsivity ranges from the near ultraviolet into the infrared spectral region, with the long wavelength cutoff at about 1100 nm due to the silicon bandgap. However, for the other optical fiber transmission windows of 1300 and 1550 nm, germanium detectors are more efficient.

Two types of photodiodes have been specially developed for optical fiber communication. They are the P-I-N and the avalanche photodiodes and their configurations are presented in Fig.1.10. One of the important requirements for higher bit-rate transmission is the speed of response of the photodetector. In the case of P-I-N photodiodes, the above-mentioned requirement is met by separating the heavily doped *n*-layer from the heavily doped *p*-layer by means of a wide high-resistance *i*-region. Thus, even for small reverse voltages, the depletion layer, and thus the electric field, extend over the entire *i*-region. The quantum efficiency of such a diode is typically about 80%; the junction capacitance of a diode designed for optical communication with an active area of 200 μm diameter amounts to 0.2 pF.

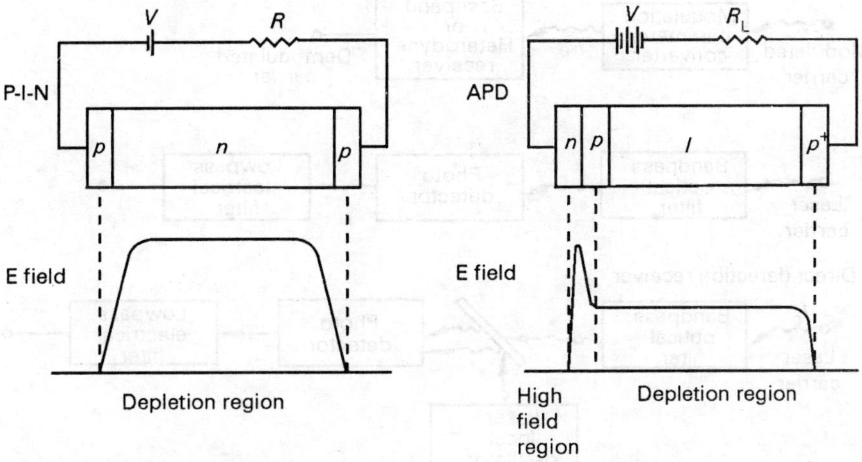

Fig. 1.10 *The P-I-N diode and avalanche photodiode.*

Due to the favourable characteristics of these silicon photodiodes, without internal gain (low dark current, short noise, small capacitance), amplifier noise generally predominates in optical receivers employing such diodes. Consequently, the receiver sensitivity can be substantially increased if photodiodes are employed in which the internal current gain resulting from avalanche multiplication of the generated charge carriers is possible. The avalanche photodiodes (APDs), however, suffer from the following disadvantages:

1. The voltage required to achieve the high-field strength for avalanche multiplication in the intrinsic layer of the diode is very high (about 1000 V).
2. The total drift times of the charge carriers is increased by the avalanche built-up process. Since the multiplication region extends over a large part of the *I*-layer, the diodes have only a very modest gain–bandwidth product. It is, therefore, necessary to restrict the multiplication to a very narrow region of thickness of about 1 μm.

The various optical receiver configurations deploying the photodetectors discussed above are depicted in Fig. 1.11. In most of the cases, the direct detection receiver is sufficient.

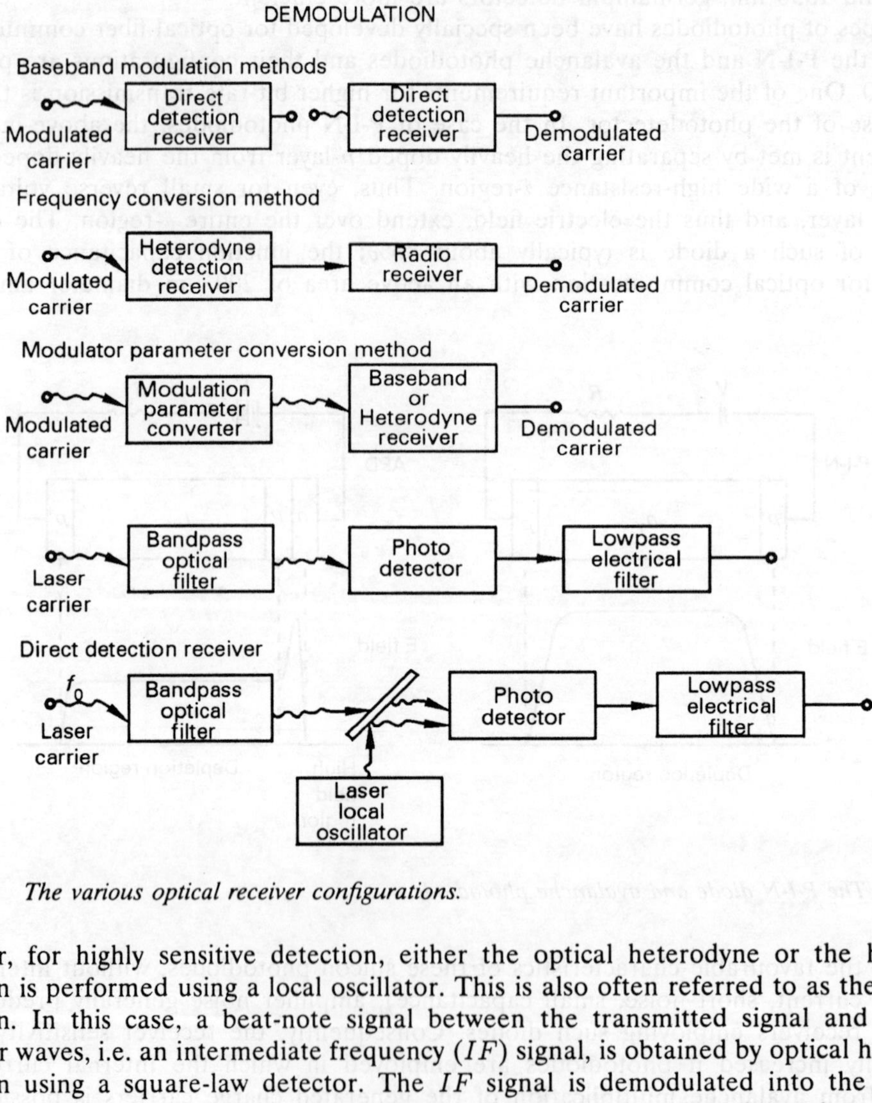

Fig. 1.11 The various optical receiver configurations.

However, for highly sensitive detection, either the optical heterodyne or the homodyne detection is performed using a local oscillator. This is also often referred to as the coherent detection. In this case, a beat-note signal between the transmitted signal and the local oscillator waves, i.e. an intermediate frequency (*IF*) signal, is obtained by optical heterodyne detection using a square-law detector. The *IF* signal is demodulated into the baseband signal through envelope, differential, or synchronous detection. In a homodyne detection system, the demodulated baseband signal is directly extracted via its optical mixing form since the local oscillator light has the same frequency as the carrier wave. An optical fiber which maintains the polarization state of the propagating signal is utilized as the transmission medium since coherent detection schemes are sensitive to the polarization states of both the transmitted signal and the local oscillator waves.

1.6 Fiber-optic communication systems

Fiber-optic communication systems are essentially designed for digital transmission, though analog transmission is adopted in certain specialized cases. In order to transmit electrical signals *via* optical waveguides, optical transmission systems are used. The principal components are the optoelectronic transducer as the optical transmitter at the transmitting end of the link, the actual optical waveguide, and the optoelectronic transducer as photodetector at the receiving end of the link (Fig. 1.12). The electrical signals at the terminals of the optoelectronic transducer at the transmitting end are converted into optical signals and the light is launched into the fiber. The light passes through the fiber as shown in the figure; at the end of the link the optical signals are converted back into electrical signals in the optoelectronic transducer and are available at the output terminals.

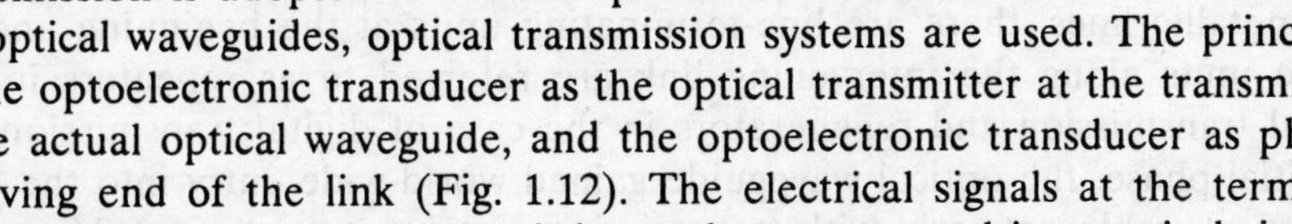

$$i_2 = P_2 \frac{\eta_e}{h\nu}$$

$$\frac{S_2}{N} = \frac{(m.i_2)^2}{i_N^2} \quad \text{signal to noise ratio}$$

e = Electron charge (1.6×10^{-19} C)

E/O = Optoelectronic transducer — Electrical to optical

η = Quantum efficiency of photodiode

h = Planck's constant (6.6×10^{-34} Ws^2)

i_1 = Transmitted signal

i_2 = Received signal

i_N = Noise current

m = Modulation factor of the optical power

ν = Frequency of optical signal

O/E = Optoelectronic transducer — Optical to electrical

P_1, P_2 = Optical power

***Fig. 1.12** A schematic diagram of the optical transmission path.*

Since in the simplest case, both ends of a line consist of two terminals carrying electrical signals, optical transmission systems can be measured and assessed in the same way as conventional electrical systems. Finally, for the user and the maintenance personnel, the optical transmission can recede completely into the background, since the link is judged solely in terms of electrical parameters.

The transmission capacity and the maximum range between the transmitter and receiver, i.e., the elementary cable section length, are generally the two system parameters of greatest importance and interest in optical communication. If the distance to be bridged is greater than the maximum section length as determined by the system, repeaters or regenerators are inserted. In these units, the received optical signal is converted into an electrical signal, amplified or regenerated and then converted back into an optical signal. (As in the classical system for metallic lines, there are line-terminating units at the beginning and the end of the link; the units along the intervening link are referred to as repeaters in the case of analog signal transmission and regenerators in the case of digital transmission).

In the initial phase, the optical waveguide gained world-wide entry into the junction and inter-office trunk cable network, with transmission systems operating at 34 and 45 Mbits/s at an optical wavelength of around 850 nm. The reason for this was that at that time, neither the available fibers nor the transmitting and receiving components were suitable for transmission at higher bit rates. Now, standard single-mode transmission systems operating at 140 and 565 Mbits/s or even higher bit rates are increasingly being introduced. Because of the obvious advantages discussed earlier, a world-wide trend towards longer wavelengths for optical transmission has since become discernible. In order to exploit the optical windows at 1300 nm and 1550 nm, a new generation of sources have to be developed and perfected. The present R&D activities are doing just that and the first commercial systems employing a wavelength of 1300 nm are already in use.

1.7 Integrated optics

Integrated optics is based on the guiding of electromagnetic energy at optical frequency by thin films for wavelengths of interest for optical communication (0.8 μm to 1.6 μm). The channel waveguide described above has a typical cross-section of 1 μm $\times$ 1 μm which is even smaller than the polysilicon lines presently used in LSI silicon circuits. The idea is to use these thin-film waveguides as basic structures of the optical devices and then to interconnect them by other thin-film waveguides to form optical circuits. If realized, these circuits would serve as miniature optical counterparts of microwave devices and networks, except that the integrated optical circuits would offer the advantages of much larger bandwidth and negligible sensitivity to interference by natural or man-made electromagnetic fields of lower frequencies.

A special role is played by semiconductors in this process, and they now appear to be very promising in promoting the goal of monolithic integrated optical circuits. For example, in lasers, it is the high-index film of the waveguides in a light-amplifying medium, and for modulators and switches, it is either an electromagnetic or electro-optic switching medium. The light wave is either amplified or switched by going through the waveguides. Naturally, the input and output terminals of the waveguides become the input and output terminals of the device. To form a laser modulator circuit, it is simple to connect the output waveguide terminal to the input waveguide terminal of the modulator. Thus, waveguides play a central role in integrated optics and for that reason, integrated optics is also called *guided wave optics*.

Lasers and detectors that are used for optical communication are AlGaAs alloy compounds grown epitaxially on GaAs substrates, or InGaAsP alloy compounds grown on InP substrates. We can thus fabricate all sorts of optical devices, passive or active, from either of the two above-mentioned alloy compounds. Integrated optical circuits formed in this way are called monolithic circuits. The above alloy compounds are good materials for transistors because of their high mobilities and saturation velocities. The possibility of forming optical devices and transistors on a common substrate allows us to combine optics with electronics in integrated circuits, thus opening up new possibilities and applications in modern electronics. We can soon expect a complete optical receiver or laser transmitter to be fabricated with all the electronical components on a small chip of GaAs or InP. They will have far-reaching consequences in automation and optical communication.

1.8 Future trends

If higher transmission capacities are required, it appears, at present and probably in the foreseeable future, that a fiber-optic system operating at 565 Mbits/s (about 4×140 Mbits/s) is more economical than several 140 Mbits/s systems employing space or wavelength-division multiplexing. Owing to the dispersion characteristics of optical waveguides, for systems operating at 565 Mbits/s or higher, only single-mode fibers, operating preferably at a wavelength of 1300 nm, are suitable. At wavelengths between 1500 and 1600 nm, it is possible to achieve even lower losses and therefore, longer repeater spacings than at 1300 nm. But, for high-speed pulse transmission, it is necessary to use an optical transmission with an extremely narrow spectral bandwidth. As an answer to this, distributed feedback (DFB) semiconductor lasers have been developed and are now commercially available. Experiments with these new generation sources operating at 1.55 μm repeaterless transmission over distances in excess of 100 km have been successful. Upgrading systems to 1.2 Gbits/s at the same repeater spacing is viable both at 1300 nm and 1500 nm, provided line-narrowed sources are employed.

Coherent optics fiber transmission has been the focus of research in recent years. In this scheme, the coherent properties of the laser light are utilized for optical communication. Here, signal transmission is achieved by using optical frequency or phase modulation instead of intensity modulation. In this scheme, a local-laser oscillator, to retrieve information at the receiver sites, is required. Optical heterodyne (where the carrier frequency f_c is different from the laser oscillator frequency f_o) or homodyne ($f_o = f_c$) was shown to significantly improve the signal-to-noise ratio (S/N) above that of direct detection. A large number of system experiments in laboratories using coherent techniques have been reported so far. These experiments, however, do not exhibit the ultimate in coherent system performance since they are based mainly on conventional optical device technologies. The ultimate goal of coherent optical fiber transmission consists of fully utilizing the coherent properties of laser light as well as low-loss and broad-bandwidth characteristics of the optical fibers themselves. Single-frequency semiconductor laser technology, Optoelectronic Integrated Circuit (OEIC) technology, and new technologies based upon these fundamental device technologies will intensify the foundation necessary for developing coherent optical fiber transmission systems.

2 Optical sources: The LASER

2.1 Historical background

The most important component of the optical communication system is perhaps the optical source, which is a generator of electromagnetic (EM) waves at optical frequencies. These EM waves act as carriers of information and are transmitted towards the receiver. The normal light (EM waves at optical frequencies), which we get so abundantly in nature from the sun, moon, stars, etc. or from an artificial source like an incandescent lamp, is a bundle of wavelengths of varying amplitudes, without any phase relationship whatsoever, as shown in Fig. 2.1(a). Such waves are unsuitable for carrying large amounts of information over large distances as was obvious from the experiment of Alexander Graham Bell when he

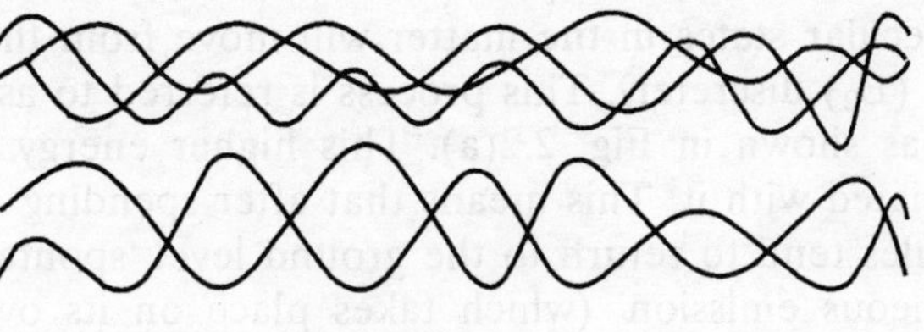

(a) Incoherent

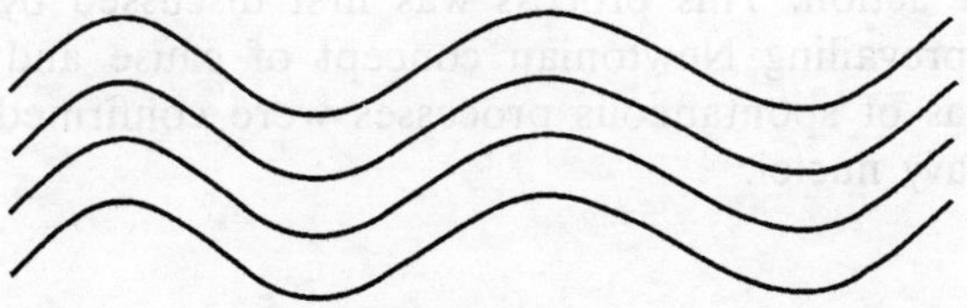

(b) Coherent

Fig. 2.1 (a) Incoherent light having no phase relationship. (b) Coherent light from a laser source.

tried to use the sun as the source. This was mainly due to the limitation in modulating such an 'incoherent' carrier with information. Thus, one had to wait till the advent of a 'coherent' EM wave generator in the form of a LASER (Light Amplification by Stimulated Emission of Radiation), which is a successor of the MASER (Microwave Amplification by Stimulated Emission of Radiation). Though there are some fundamental differences between the two devices, the basic principles underlining their functioning are the same.

While maser acts like a low-noise microwave pre-amplifier, laser is essentially an oscillator with a built-in optical resonator cavity. Maser was discovered by Townes in 1954 and laid the quantum electronic foundations for the advent of laser.

2.2 Laser fundamentals

When the electromagnetic radiation passes through a laser grade material, three basic physical processes can take place depending upon the nature of the electromagnetic radiation and the level of the laser material which is interacting with it. These three processes (depicted in Fig. 2.2) are

(a) absorption
(b) spontaneous emission
(c) stimulated emission

The simplified two-level energy level diagram of matter (atoms/molecules) before and after the interaction with electromagnetic radiation is illustrated in Fig. 2.2. The ground state of the matter will continue to exist in that level unless it is disturbed by an external agency like an appropriate electromagnetic energy (E) having a frequency f impinging on it. As per the quantum theory of matter, $E = hf$, where h is the Planck's constant. Utilizing their energy, the atomic/molecular states in the matter will move from the ground level (E_1) to the higher energy level (E_2) discretely. This process is referred to as stimulated absorption or simply 'absorption' as shown in Fig. 2.2(a). This higher energy level (E_2) will have a certain lifetime τ associated with it. This means that after spending a time τ in this energy state, the atoms/molecules tend to return to the ground level 'spontaneously'. This process, referred to as 'spontaneous emission' (which takes place on its own, without the aid of any external agencies), leads to the emission of a photon whose frequency is $(E_2 - E_1)/h$ [Fig. 2.2(b)]. This process becomes more and more probable at higher frequencies and is very crucial for laser action. This process was first discussed by Einstein in 1917 and contradicted the then prevailing Newtonian concept of cause and effect. Decades later, however, Einstein's ideas of spontaneous processes were confirmed with the discovery of radioactive decay in heavy nuclei.

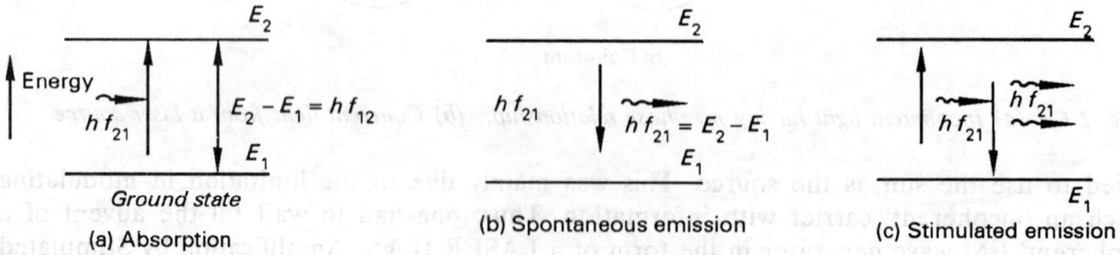

Fig. 2.2 (a) The process of stimulated absorption. (b) The spontaneous emission process. (c) The stimulated emission process.

The third and final process, i.e., the stimulated emission, ideally follows the classical physical concepts and is shown in Fig. 2.2(c). If a photon interacts with matter which is in an excited state, it acts as a 'stimulant' to cause the transition from the higher energy state to the ground level. In such a case, the differential energy released due to the transition from the higher energy state E_2 to the ground level E_1 will be added to the stimulating photon in phase, leading to amplification. This process of quantum amplification was utilized in the case of ammonia maser successfully. Such a process will lead to 'coherence' and light amplification in the case of a laser.

2.3 Einstein's A and B coefficients

Assuming that the system is in a state of thermal equilibrium, the distribution of energy states follows the well-known Boltzmann's law as shown in Fig. 2.3(a). If the number of

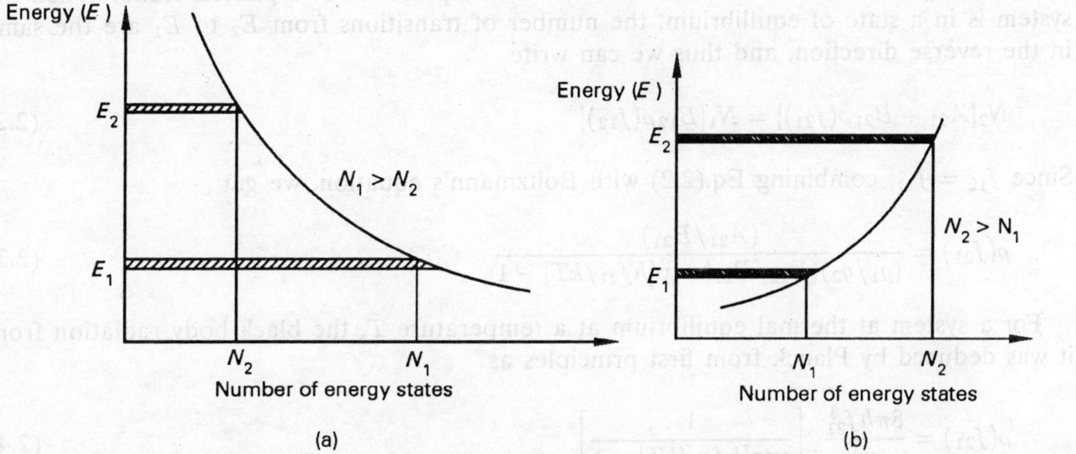

Fig. 2.3 The distribution of energy states for (a) normal Boltzmann distribution ($N_1 > N_2$) and (b) Population inversion ($N_2 > N_1$).

electronic atomic/molecular states associated with the energy levels E_1 and E_2 are N_1 and N_2 respectively, the Boltzmann theory states that

$$\frac{N_2}{N_1} = \frac{g_2 \exp\{-E_2/kT\}}{g_1 \exp\{-E_1/kT\}} = \frac{g_2}{g_1} \exp\left[\frac{-(E_2 - E_1)}{kT}\right] \tag{2.1}$$

where g_1 and g_2 are referred to as degeneracy factors indicating the number of states available at the energy levels E_1 and E_2 for possible transitions. Einstein defined the A and B radiative rate coefficients when the system is in radiative equilibrium, as follows:

(i) The spontaneous emission coefficient A_{21} is the probability per second that the system at energy level E_2 will make a transition to a state at E_1 without any external influence. No external photon is required. By the same token, A_{12}, which represents the probability transition from a ground state (E_1) to an excited state (E_2) spontaneously, is non-existent.

(ii) The stimulated emission coefficients B_{12} and B_{21} are caused by the incoming electromagnetic radiation and have to be defined in terms of it. They are then the probability per second that the system will make a transition in the presence of a photon cloud of energy density $\rho(f)\, df$, where df is the frequency range of the photon cloud.

Assuming that the system is in the ground state E_1 and it makes a transition, the probability per second that it will make a transition to the upper state E_2 by absorbing a quantum of radiation hf_{12} is $B_{12}\rho(f_{12})$. Similarly, if the system is already in the upper state E_2 and interacts with a quantum of radiation hf_{21}, then the probability per second that it will make a transition to the ground state E_1 is $[A_{21} + B_{21}\rho(f_{21})]$ where the first term corresponds to the spontaneous emission in the absence of a photon cloud and the second term corresponds to the stimulated emission in the presence of a photon cloud. When the system is in a state of equilibrium, the number of transitions from E_2 to E_1 are the same in the reverse direction, and thus we can write

$$N_2[A_{21} + B_{21}\rho(f_{21})] = N_1[B_{12}\rho(f_{12})] \tag{2.2}$$

Since $f_{12} = f_{21}$, combining Eq.(2.2) with Boltzmann's equation, we get

$$\rho(f_{21}) = \frac{(A_{21}/B_{21})}{(g_1/g_2)(B_{12}/B_{21})\exp[hf_{21}/kT] - 1} \tag{2.3}$$

For a system at thermal equilibrium at a temperature T, the black body radiation from it was deduced by Planck from first principles as

$$\rho(f_{21}) = \frac{8\pi h f_{21}^3}{c^3}\left[\frac{1}{\exp[hf_{21}/kT] - 1}\right] \tag{2.4}$$

Comparing the two relations given in Eqs.(2.3) and (2.4), we get

$$\frac{A_{21}}{B_{21}} = \frac{8\pi h f_{21}^3}{c^3} = \frac{8\pi h}{\lambda^3} \tag{2.5}$$

and $$g_1 B_{12} = g_2 B_{21} \tag{2.6}$$

For a system with negligible degeneracy, we have

$$B_{12} \approx B_{21} = B \tag{2.7}$$

and $$A_{21} = \frac{8\pi h f_{21}^3 B}{c^3} \tag{2.8}$$

By the very definition, the lifetime τ_s of a system can be related to A_{21} as:

$$A_{21} = 1/\tau_s \tag{2.9}$$

Example 2.1 Obtain the ratio of the stimulated emission rate to the spontaneous emission rate of white light source operating at 900 K, whose average emission wavelength is 0.6 μm.

Solution The average operating frequency is given by

$$f = \frac{c}{\lambda} = \frac{3 \times 10^8}{0.6 \times 10^{-6}} = 5 \times 10^{14} \text{ Hz}$$

We have seen that

$$\frac{\text{stimulated emission rate}}{\text{spontaneous emission rate}} = \frac{B_{21}\rho(f)}{A_{21}} = \frac{1}{\exp(hf/kT) - 1}$$

$$= \frac{1}{\exp\left[\dfrac{6.626 \times 10^{-34} \text{ J.s} \times 5 \times 10^{14}\text{Hz}}{1.381 \times 10^{-23} \text{ J/K} \times 900 \text{ K}} - 1\right]^{-1}}$$

$$= \exp(-26.66) \sim 3 \times 10^{-13}$$

The result obtained in Example 2.1 clearly shows that for a system in thermal equilibrium emitting optical radiation, spontaneous emission is predominant. This clearly explains why the optical radiation emitted by natural/artificial sources is 'random' or 'incoherent' in nature. It is, therefore, obvious that in order to obtain 'coherent' radiation, we should enhance the stimulated emission rate and make it greater than the spontaneous emission rate. Besides, in order to ensure that the system is a net emitter of radiation, we should see that the population density N_2 in the upper energy state E_2 is larger than the population density N_1 in the lower energy state E_1.

2.4 Population inversion and gain medium

According to Boltzmann's law, the population N_2 at the energy level E_2 will always be less than the population N_1 at the energy level E_1. In other words, it means that a medium will always be a net absorber of electromagnetic radiation under thermal equilibrium conditions. In order to make the medium an amplifying medium to achieve optical amplification by stimulated emission of radiation, one has to create a situation where the population N_2 at energy level E_2 is more than the population N_1 at energy level E_1. This is normally referred to as 'population inversion' and is shown in Fig. 2.3(b).

This kind of 'anti-Boltzmann distribution' can be achieved artificially under non-equilibrium conditions by injecting energy into the medium (by sustaining more number of states at the energy level E_2 than what will be sustained at the energy level E_1). This population inversion can only be created and maintained by 'optical pumping' whereby the medium acts like a net amplifier of radiation. However, in nature, the situation at energy levels corresponding to optical frequencies will not be as simple as described above, and one has to deal with three-level or four-level energy systems in practical laser systems. The growth of optical radiation in an absorbing medium ($N_2 < N_1$) and in an amplifying medium ($N_2 > N_1$) is illustrated in Figs. 2.4(a) and (b).

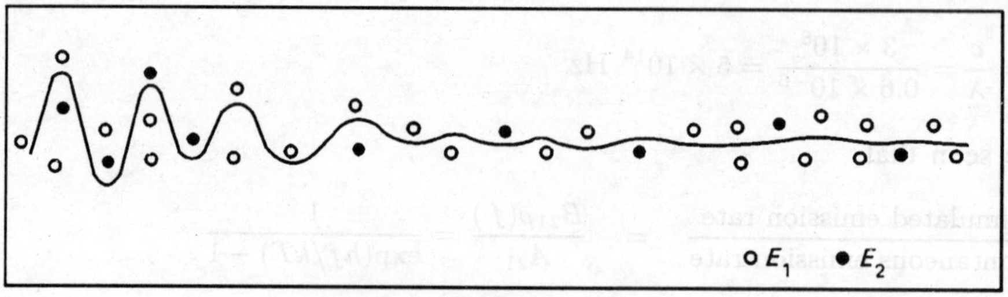

(a) Loss medium ($N_1 > N_2$)

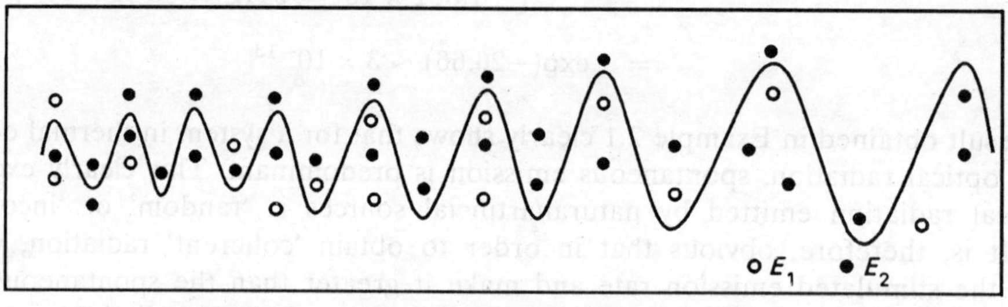

(b) Gain medium ($N_2 > N_1$)

Fig. 2.4 The radiation of the transition frequency for (a) absorbing medium ($N_1 > N_2$) and (b) amplifying medium ($N_2 > N_1$).

Example 2.2 An atom has two energy states with a transition wavelength $\lambda = 694.3$ nm. What percentage of the atoms exists in the higher energy state at room temperature (T = 300 K)? Assume that all the atoms are in one or the other of these two energy levels.

Solution

$$E_2 - E_1 = \frac{hc}{\lambda} = \frac{6.626 \times 10^{-34}\,\text{J}\cdot\text{s} \times 3 \times 10^8\ \text{m/s}}{694.3 \times 10^{-9}\ \text{m}}$$

$$= 2.863 \times 10^{-19}\ \text{J}$$

From Boltzmann's law,

$$\frac{N_2}{N_1} = \exp\left[\frac{-(E_2 - E_1)}{kT}\right]$$
$$= \exp\left[-\frac{2.863 \times 10^{-19}\text{J}}{1.381 \times 10^{-23}\text{J/K} \times 300\text{K}}\right]$$
$$= \exp(-69.105) = 9.728 \times 10^{-31}$$

From this, it can be seen that at room temperature (300 K), the number of atoms in the higher energy state are negligible and they have to be created by optical pumping in order to achieve population inversion.

The effective spectral radiation energy density associated with the light beam is related to the spectral irradiance of the beam by $\rho(f) = I(f)/c$. Thus, the stimulated absorption rate = $N_1 I(f)/c \cdot B_{12}$ and the stimulated emission rate = $N_2 I(f)/c \cdot B_{21}$.

Since the system is not in a state of radiative equilibrium due to optical pumping, the number of energy states found at these two levels will be a function of the time and can be written as

$$-\frac{dn_{12}}{dt} = N_1 \frac{I(z)}{c} B_{12} - N_2 \frac{I(z)}{c} B_{21} \tag{2.10}$$

$$= B(N_1 - N_2)\frac{I(z)}{c} \tag{2.11}$$

since $B_{12} = B_{21} = B$. The contribution due to a spontaneous emission process is ignored for the time being. We can also express the rate of photon beam changes in terms of beam intensity changes at z and $z + dz$ as

$$-\frac{dn_{12}}{dt} = [I(z) - I(z + dz)] \; \frac{\text{area}}{hf_{12}} \tag{2.12}$$

where $I(z)$ and $I(z + dz)$ are the photon beam intensities at z and $z + dz$ respectively, assuming a uniform cross-sectional 'area' for the beam which is passing through the medium. We can then combine Eqs.(2.11) and (2.12) as

$$[I(z) - I(z + dz)]\frac{\text{area}}{hf_{12}} = B(N_1 - N_2)\frac{I(z)}{c} \tag{2.13}$$

If dz is sufficiently small, this can be written as

$$\frac{-dI(z)\Delta z}{dz}\,\frac{\text{area}}{hf_{12}} = B(N_1 - N_2)\frac{I(z)}{c} \tag{2.14}$$

which can be expressed in the form

$$\frac{dI(z)}{dz} = \frac{B(N_1 - N_2)I(z)}{\text{volume} \cdot c} hf_{12} \tag{2.15}$$

We can easily solve the equation for $I(z)$ as a function of the distance through which the light travels in the medium, with the result

$$I(z) = I(0)\exp(-\alpha z) \tag{2.16}$$

where $$\alpha = \frac{B(N_1 - N_2)}{\text{volume}} \cdot \frac{hf_{12}}{c} \tag{2.17}$$

If population inversion exists in the medium, it acts like a 'gain medium' where the gain coefficient β is given by

$$\beta = -\alpha = \frac{B(N_2 - N_1)}{\text{volume}} \cdot \frac{hf_{12}}{c} \tag{2.18}$$

The light beam is then amplified instead of being attenuated. The process causing this growth in beam irradiance, as illustrated in Fig. 2.4(b), is the result of light amplification by stimulated emission and thus is the underlying basis for laser operation.

Example 2.3 If the radiance of light doubles on passing through a laser amplifier of one metre of gain medium, calculate the gain coefficient β of the medium.

Solution $I(z) = I(0)\exp(\beta z)$

Here $z = 1$ m. Therefore,

$$\frac{I(1)}{I(0)} = 2 = \exp(\beta)$$

Taking natural logarithms on both sides, we get

$$\beta = \log_e 2 = 0.693$$

2.5 Optical feedback and laser oscillation

Since we are interested in a carrier at optical frequencies to act as a transmitter in the optical communication system, we have to organize feedback in the optical amplifier in order to create laser oscillations. This is analogous to an electronic oscillator where a positive feedback is organized from an LC resonator circuit to achieve sustained oscillations at RF frequencies. This concept has to be extended to optical frequencies which has resulted in the development of an optical resonator cavity as shown in Fig. 2.5.

The above structure acts like a Fabry–Perot etalon. Though the gain of the medium is small for a single pass, it can be substantial after multiple passes due to the feedback. After sufficient energy is built up, the laser quality (coherent) radiation emerges through the partially reflecting mirror. A stable output can result when the overall gain of the system is larger than the sum total of the different losses like absorption, scattering, diffraction, etc.

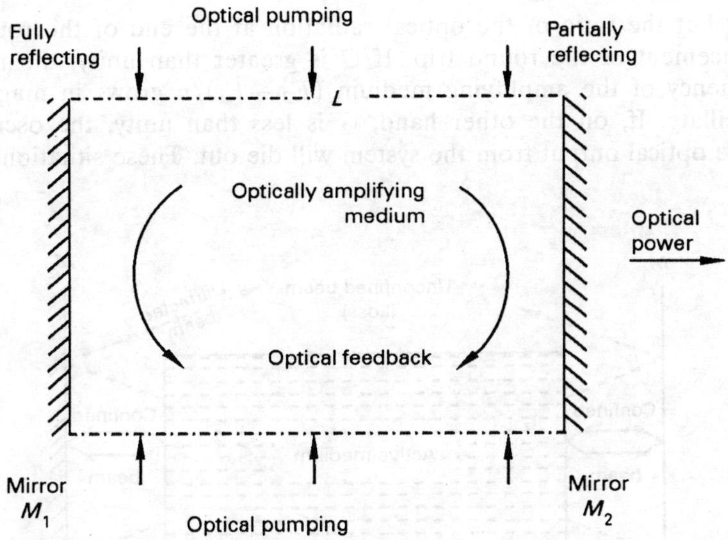

Fig. 2.5 *Optical resonator.*

By placing the amplifying medium in between the parabolically reflecting mirrors that direct the optical beam back and forth through the gain medium several times, the effective length of the amplifying medium becomes many times more than its actual length.

For example, using a combination of a 100% reflecting flat mirror and a 98% reflecting flat mirror aligned parallelly, the effective length of the amplifying medium would roughly be 50 times the actual distance between the two mirrors. Thus, though the single pass amplification is quite small for the population-inverted amplifying medium, the total amplification provided by multiple passes would be substantial.

In theory, an amplifying medium will become a gain medium when $N_2 > N_1$. However, in practice, it is not sufficient that N_2 is simply greater than N_1, if the system has to break into sustained oscillations. The sustained oscillations in the system are possible only if the combined losses in the system are overcome by a gain which is referred to as threshold gain or β_{th}. While travelling from mirror M_1 to mirror M_2 through the amplifying medium, the beam irradiance will increase from its initial value I_0 to a value I given by

$$I = I_0 \exp[(\beta - \alpha_e)L] \tag{2.19}$$

where L is the distance between the mirrors, β is the gain coefficient, and α_e is the distributed loss coefficient of the system. After reflection from the mirror M_2, the relative beam irradiance will become $R_2 \exp[(\beta - \alpha_e)L]$, where R_2 is the reflectivity of the mirror M_2 which is the ratio of the reflected to the incident optical radiation. After a return trip and subsequent reflection from mirror M_1 whose reflectivity is R_1, the final round-trip gain of the optical radiation is given by

$$G = R_1 R_2 \exp[2(\beta - \alpha_e)L] \tag{2.20}$$

This is nothing but the ratio of the optical radiation at the end of the round trip to that at the commencement of the round trip. If G is greater than unity, the radiation at the transition frequency of the amplifying medium $(E_2 - E_1)/n$ grows in magnitude and the system will oscillate. If, on the other hand, G is less than unity, the oscillations will be damped and the optical output from the system will die out. These situations are illustrated in Fig. 2.6.

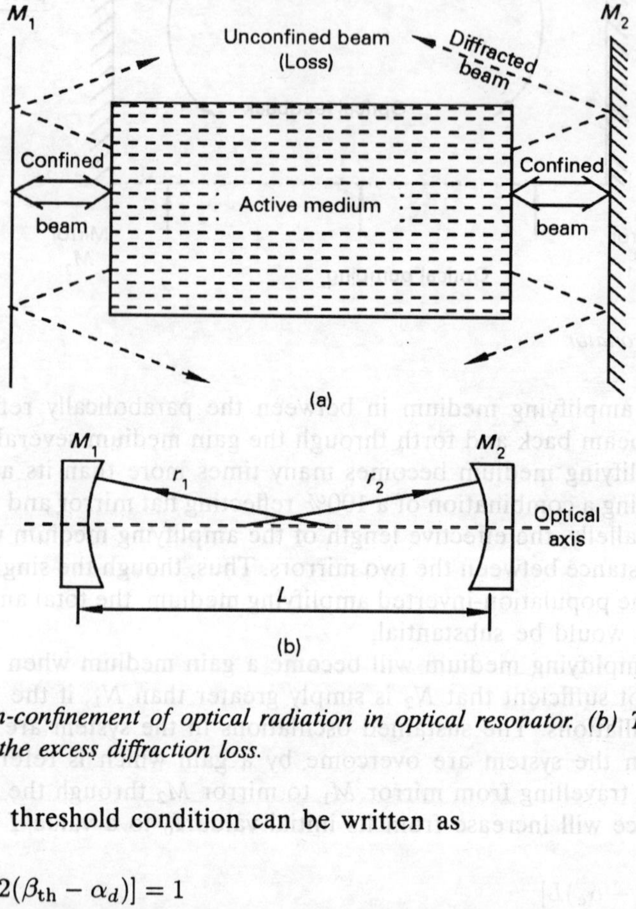

Fig. 2.6 (a) The non-confinement of optical radiation in optical resonator. (b) The optimum optical resonator to overcome the excess diffraction loss.

From Eq.(2.20), the threshold condition can be written as

$$G = R_1 R_2 \exp[2(\beta_{\text{th}} - \alpha_d)] = 1 \tag{2.21}$$

from which it can be written that

$$\beta_{\text{th}} = \alpha_d + \frac{1}{2L} \ln(1/R_1 R_2) \tag{2.22}$$

The second term on the right-hand side of the equation represents the optical output from the laser system since the gain is equal to the sum of total losses in the system.

2.6 Optical resonator: The stability condition

So far, it has been assumed that two plane parallel mirrors can constitute an optical resonator. However, this is not an optimum solution since in such a situation the diffraction losses may be considerable due to the non-confinement of optical radiation at the edges due to flat mirrors as shown in Fig. 2.6(a). The unconfined optical radiation, which can be considerable, depending upon the system configuration, will not contribute to the stimulated emission process and can make all the difference between lasing and non-lasing. Thus, it is necessary to engineer an optical resonator cavity where the excess diffraction losses are overcome by curving the mirrors sufficiently inward towards the cavity. The necessary theory for designing an optimum optical resonator cavity was found by Fox and Li in 1964 in terms of system parameters as shown in Fig. 2.6(b). From their theoretical analysis, the stability condition for a low-loss optical resonator configuration is given by

$$0 \leq g_1 g_2 \leq 1 \tag{2.23}$$

where $g_1 = 1 - \dfrac{L}{r_1}$ and $g_2 = 1 - \dfrac{L}{r_2}$ (2.24)

Example 2.4 In a cartesian coordinate system, let g_1 be the abscissa and g_2 be the ordinate. Plot the boundary curves and shade those regions associated with g_1, g_2 satisfying the stability condition.

Solution The stability condition given by $0 \leq g_1 g_2 \leq 1$ can be subdivided into three distinct regions:

$0 < g_1 g_2 < 1$ most stable

$g_1 g_2 = 0$; $g_1 g_2 = 1$ marginally stable

$g_1 g_2 < 0$; $g_1 g_2 > 1$ unstable.

A thin line separates the stable and unstable conditions which can be seen from Fig. 2.7 which are generated from the above conditions. This figure is referred to as the 'stability diagram' and can be used in designing an optical resonator cavity.

Example 2.5 In an optical resonator cavity, consider two mirrors with radii of curvature $r_1 = +60$ cm and $r_2 = +120$ cm. Calculate:

(a) the refraction for which they would form a marginally stable configuration,
(b) the range of separations for which they would be stable, and
(c) the range of separations for which they would be unstable.

Solution (a) The conditions for a marginally stable configuration are given by $g_1 g_2 = 0$ and $g_1 g_2 = 1$.

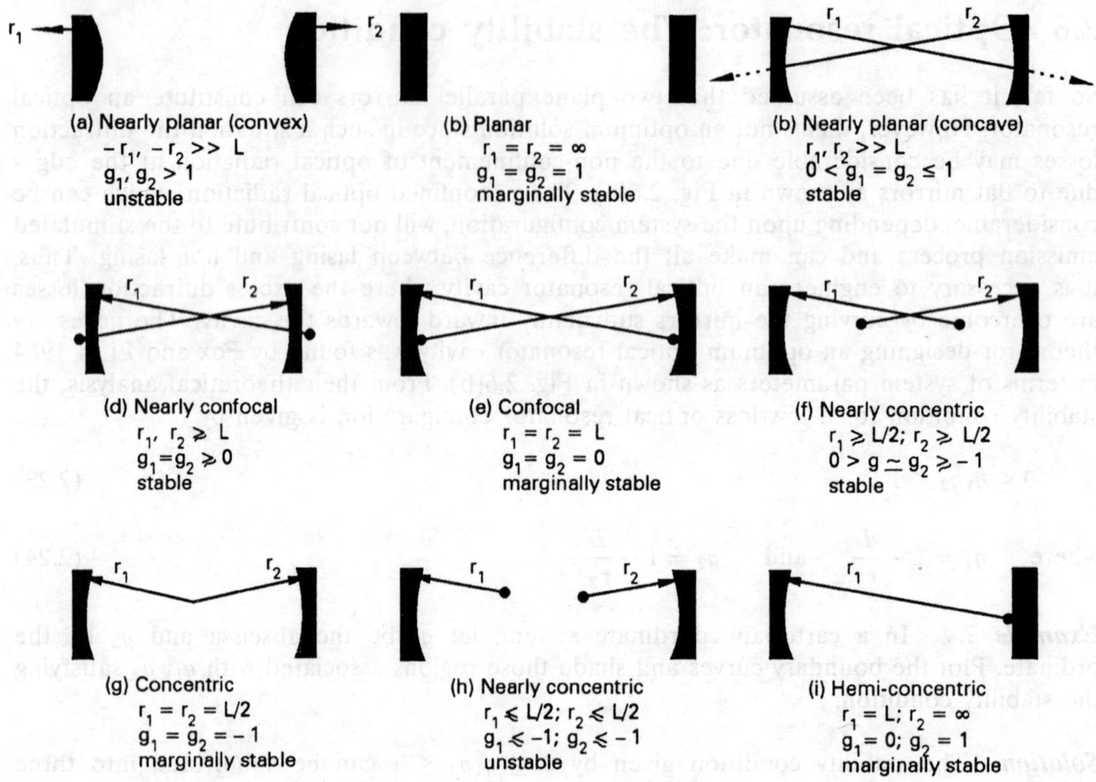

Fig. 2.7 Fox and Li condition for optical resonator stability.

For $g_1 g_2 = 0$, we have

$$g_1 g_2 = \left[1 - \frac{L}{r_1}\right]\left[1 - \frac{L}{r_2}\right] = 0$$

or $$1 - \frac{L}{r_1} - \frac{L}{r_2} + \frac{L^2}{r_1 r_2} = 0$$

or $$\frac{L}{r_1} + \frac{L}{r_2} - \frac{L^2}{r_1 r_2} = 1$$

or $$\frac{L}{60} + \frac{L}{120} - \frac{L^2}{60 \times 120} = 1$$

Multiplying throughout by 60 × 120, we get

$$120L + 60L - L^2 = 60 \times 120$$
$$L^2 - 180L + 60 \times 120 = 0$$

Therefore, $L = 60$ cm, 120 cm

Similarly for $g_1 g_2 = 1$, we have

$$\left[1 - \frac{L}{60}\right]\left[1 - \frac{L}{120}\right] = 1$$

or $$\frac{L}{60} + \frac{L}{120} - \frac{L^2}{60 \times 120} = 0$$

or $$L^2 - 180L = 0$$

$$L(L - 180) = 0$$

This is satisfied for $L = 0$ which is not practical since there is no space for laser medium. It is also satisfied for $L = 180$ cm.

The emergence of the optical resonator cavity from those at lower frequencies is depicted in Fig. 2.8. From this figure it can be seen that the optical resonator cavity is totally different from an LC circuit at HF frequencies. Thus, it has taken almost two years for the optical resonator cavity to be conceived and developed.

2.7 Laser system

The outline of a practical laser system has emerged by incorporating the concepts which we have discussed so far. It should therefore consist of the following subsystems:

(a) A suitable material whose energy-level diagram and the corresponding lifetimes are suitable for the lasing action.
(b) A means for creating a population-inversion in this material, thus making it an amplifying medium.
(c) An appropriate optical resonator cavity leading to oscillations at optical frequencies resulting in the laser output.

The laser materials could be in any physical state namely solids, gases, liquids, etc., the only criterion being a suitable energy-level diagram which can result in the lasing action. We will now discuss two of the early laser systems which correspond to the three-level and four-level systems representing the solid state and gas laser system.

The typical energy-level diagrams of ruby (solid state) and Helium–Neon (gas) which correspond to the three-level and four-level laser systems are given in Fig. 2.9. These were the first two lasers to have been developed and fabricated. As can be seen from the figure, the population inversion ($N_2 > N_1$) is created in the ruby system by a combination of optical pumping and fast decay of level 3. Then the ruby crystal becomes a gain/amplifying medium for optical frequencies corresponding to the frequency $(E_2 - E_1)/h$. By a similar combination of events, (He–Ne) gas becomes a gain/amplifying medium at the frequency

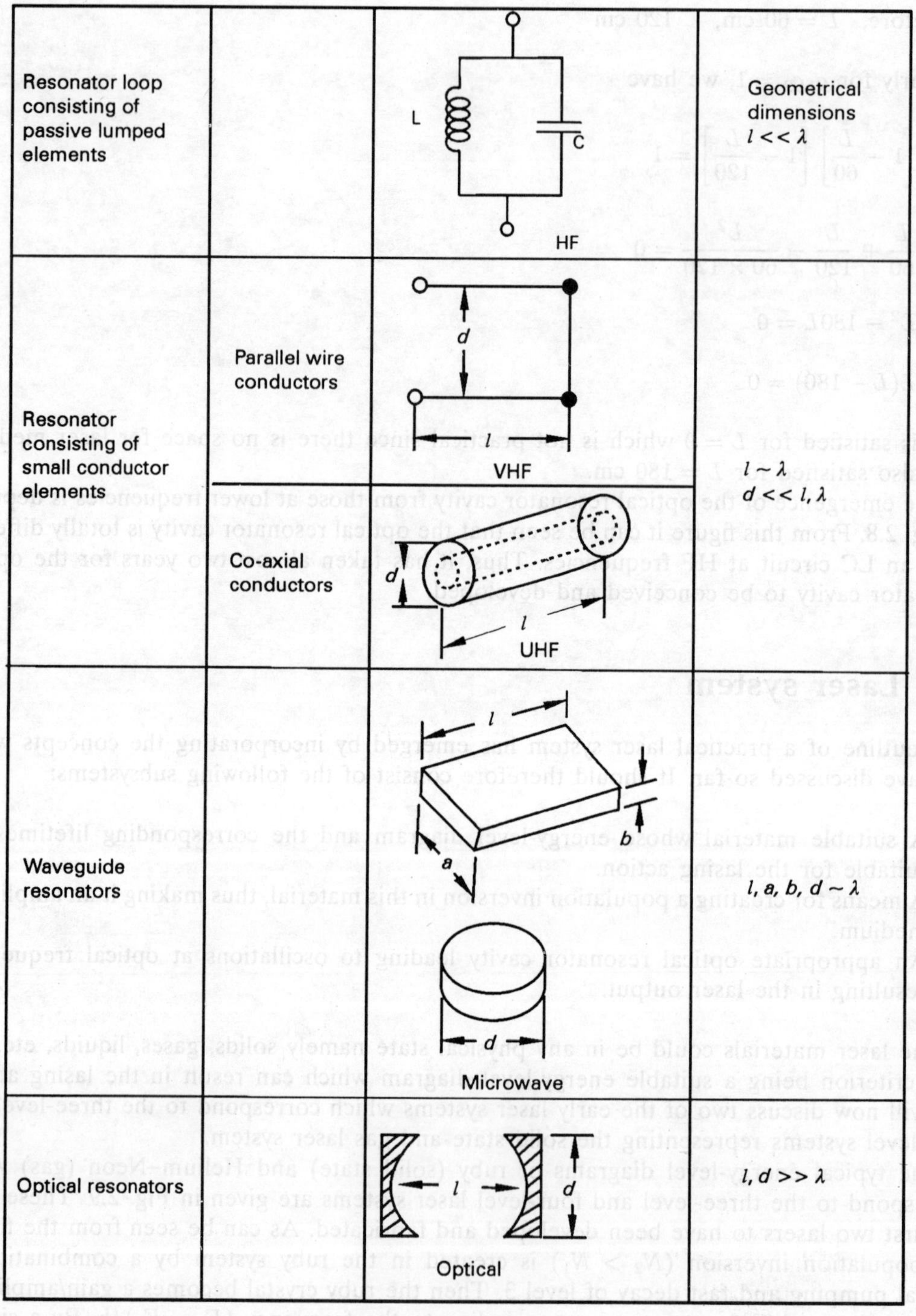

Fig. 2.8 The emergence of an optical resonator cavity.

$(E_3 - E_2)/h$, where population inversion is created between the energy levels E_3 and E_2. The fabricational details of these two lasers are given in Fig. 2.10. In Fig. 2.10(a), the ruby crystal in the form of a cylindrical rod is kept inside a helical flash lamp which provides the optical pumping energy. The xenon flash lamp is energized by charging a capacitor with the help of HV supply by closing the switch S_2 just as a camera flash is operated. The intense optical radiation emanated by the flash lamp gets absorbed by the ruby crystal, which leads to the optical inversion. The two mirrors M_1 and M_2 are fully reflecting and the optical resonator is partially reflecting. Though the initial photons are generated by spontaneous emission due to the transition from E_2 to E_1, due to successive reflections from the two mirrors which are parallel to each other, stimulated emission predominates. The laser radiation which finally emerges through the partially reflecting mirror at a wavelength of 694.8 nm possesses the following three vital characteristics which are crucial for optical communication:

(a) Directionality
(b) Coherence
(c) Spectral purity.

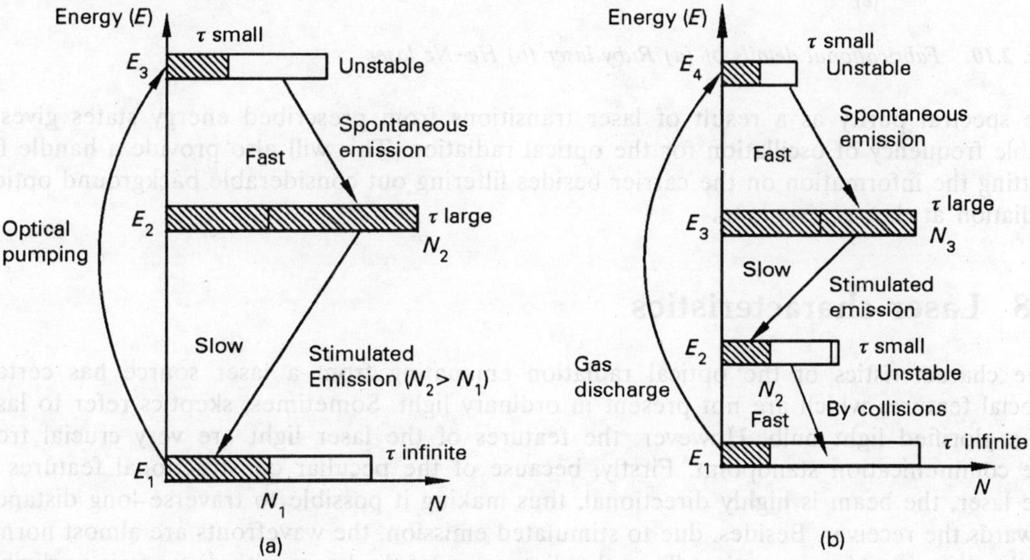

Fig. 2.9 Typical energy level diagram for lasing action.

The fabricational details of a He–Ne laser are given in Fig. 2.10(b). The He–Ne laser is usually used at 632.8 nm since it is in the visible region of the electromagnetic spectrum.

The directionality property of the laser radiation leads to the concentration of the optical energy into small areas thus providing high optical intensities. This is very useful for traversing long distances to reach the receiver in an optical communication link. Secondly, the coherence property, which is a result of predominating stimulated emission process, will be an important handler for modulating (phase) this carrier with information. Finally,

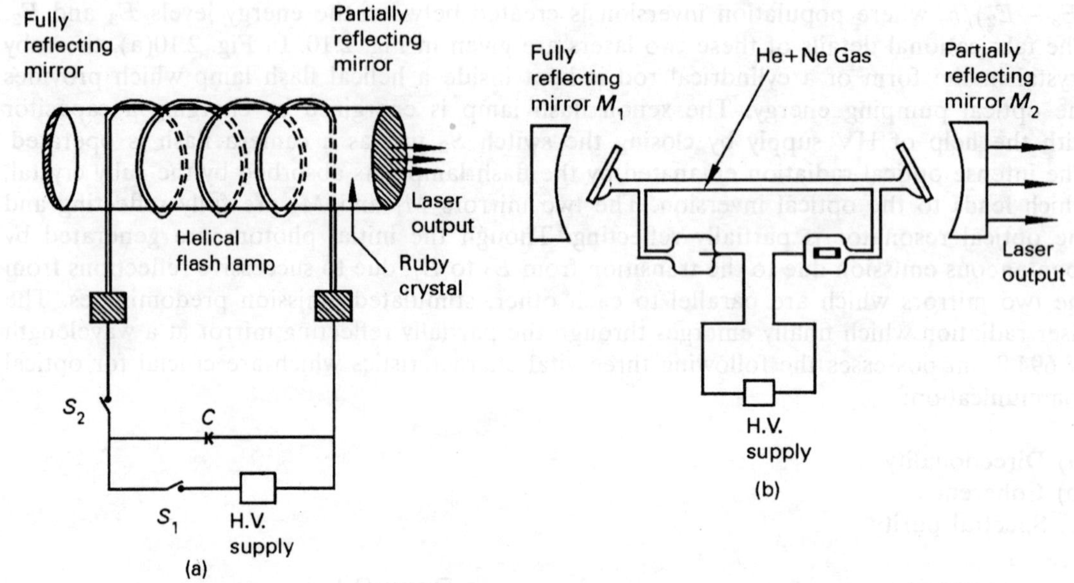

Fig. 2.10 Fabricational details of (a) Ruby laser (b) He–Ne laser.

the spectral purity as a result of laser transitions from prescribed energy states gives a stable frequency of oscillation for the optical radiation. This will also provide a handle for putting the information on the carrier besides filtering out considerable background optical radiation at the receiver end.

2.8 Laser characteristics

The characteristics of the optical radiation emanating from a laser source has certain special features which are not present in ordinary light. Sometimes, skeptics refer to laser as a glorified light bulb. However, the features of the laser light are very crucial from the communication standpoint. Firstly, because of the peculiar constructional features of the laser, the beam is highly directional, thus making it possible to traverse long distances towards the receiver. Besides, due to stimulated emission, the wavefronts are almost normal to the direction of propagation. Thus the divergence of the beam with distance is negligible. This makes it possible to collect the optical energy at the receiver end with small optical antennas. Besides, the phase coherence of the monochromatic radiation leads to energy densities inside the laser beam to be much higher than that of ordinary light. The spectral purity and coherence make it a suitable candidate for frequency and phase modulation besides the usual intensity modulation. The easy polarizability of optical radiation makes it possible to consider the polarization modulation schemes as well.

Though the spectral purity of the laser light is much higher than that of ordinary light, the linewidth is broad because the energy levels between which the laser transition takes place are themselves broad. This leads to a relatively broad amplification of the gain medium as shown in Fig.2.11(a). However, the optical resonator cavity can sustain the largest (theoretically infinite) number of modes given by the frequency

$$f = \frac{qc}{2nL} \tag{2.25}$$

where c is the velocity of light in vacuum, n is the refractive index of the medium between the two mirrors, L is the distance of separation of the two mirrors and q is an integer.

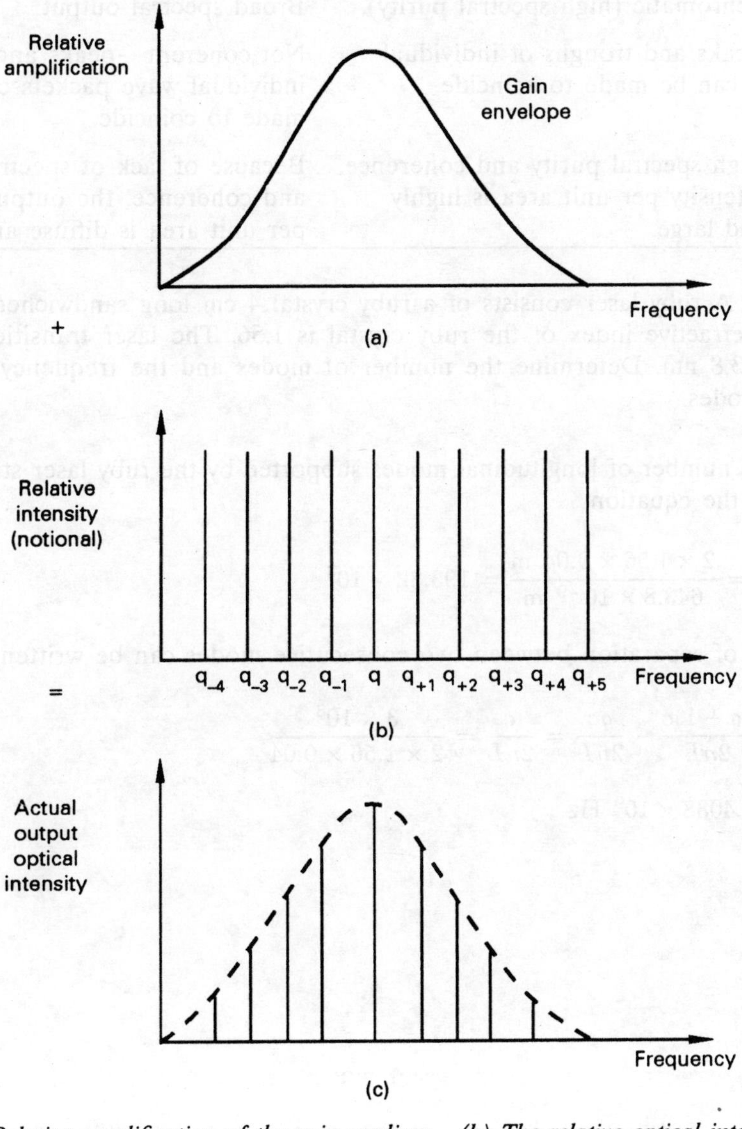

Fig. 2.11 (a) Relative amplification of the gain medium. (b) The relative optical intensity distribution as a function of the frequency. (c) The actual optical intensity distribution as a function of the frequency.

This leads to the optical intensity distribution as a function of frequency as shown in Fig.2.11(b). A combination of the laser amplifying medium and the optical resonator cavity will result in a frequency distribution of the optical output from the laser as shown in Fig.2.11(c). A comparison of the characteristics of laser light and ordinary light is given in Table 2.1.

Table 2.1 Comparison of the characteristics of laser and ordinary light.

Laser light	Ordinary light
Highly monochromatic (high spectral purity).	Broad spectral output.
Coherent—peaks and troughs of individual wave packets can be made to coincide.	Not coherent—peaks and troughs of individual wave packets cannot be made to coincide.
Because of high spectral purity and coherence, the output intensity per unit area is highly directional and large.	Because of lack of spectral purity and coherence, the output intensity per unit area is diffuse and small.

Example 2.6 A ruby laser consists of a ruby crystal 4 cm long sandwiched between two mirrors. The refractive index of the ruby crystal is 1.56. The laser transition has a peak emission at 643.8 nm. Determine the number of modes and the frequency of separation between the modes.

Solution The number of longitudinal modes supported by the ruby laser structure can be obtained from the equation

$$q = \frac{2nL}{\lambda} = \frac{2 \times 1.56 \times 0.04 \text{ m}}{643.8 \times 10^{-9} \text{ m}} = 193.82 \times 10^3$$

The frequency of separation between two consecutive modes can be written as

$$f = \frac{(q+1)c}{2nL} - \frac{qc}{2nL} = \frac{c}{2nL} = \frac{3 \times 10^8}{2 \times 1.56 \times 0.04}$$

$$= 2.4038 \times 10^9 \text{ Hz}.$$

3 Optical sources: The semiconductor laser diode and light emitting diode

3.1 Historical background

The first generation of lasers to appear on the scene during the early sixties were the Ruby Laser (solid state) designed and fabricated by Theodore Maiman in November 1960 at the Hughes Aircraft Company and the He–Ne Laser (gas counterpart) by Ali Jawan in February 1961 at the Bell Laboratories. The ruby laser gives a pulsed output because of the optical pumping mechanism which charges and discharges periodically, whereas the He–Ne laser can operate in CW mode because of continuous electrical discharges through the gas, causing optical pumping. Communication engineering communities around the world started exploring this new 'optical transmitter' for carrying information in an optical communication system. They encountered serious difficulties in modulating these lasers directly. Since the pumping mechanism for these lasers was capacitive in nature, the speed of modulation was essentially limited to the rate at which the capacitor could be charged and discharged. This process being slow, only low frequency direct modulation was possible. This did not measure up to the expected bandwidth of an optical communication system. Attempts to design an external modulator also encountered serious technical difficulties. Soon the euphoria generated about the optical communication system with a large channel capacity with laser as the optical transmitter evaporated. Many researchers soon jumped out of the bandwagon of 'optical communication' as quickly as they had jumped in.

At this stage, the semiconductor laser diode appeared on the scene, reviving hopes of a compact, rugged, efficient and directly modulatable optical source. This optical source proved to be an ideal optical transmitter in a practical optical communication system which was easily deployable in the field. The basic idea of a semiconductor p–n junction emitting laser quality radiation was postulated by Bosov in 1959. However, serious work on such a device was started at the IBM Thomas J. Watson Research Centre, culminating in the development of a laboratory model semiconductor laser diode in 1964 by Nathan and his group. This work seems to have been motivated for developing a contactless high-speed printout on a photosensitive paper which could possibly speed up the computer operations considerably. However, the first device developed by this group remained in the laboratory stage only because of operational difficulties. Further years of research and development work had to be put in by Hayashi, Nathan and co-workers before a field model of the semiconductor laser diode could be fabricated in 1970.

This, along with the parallel developments in other areas, led to a revival of interest in optical communication, which ultimately culminated in the development of high-bit rate optical communication systems in the field.

3.2 Radiative recombination in semiconductor materials

All semiconductor materials can be broadly classified into direct bandgap and indirect bandgap materials as shown in Fig. 3.1. The Y-axis represents the electron energy while

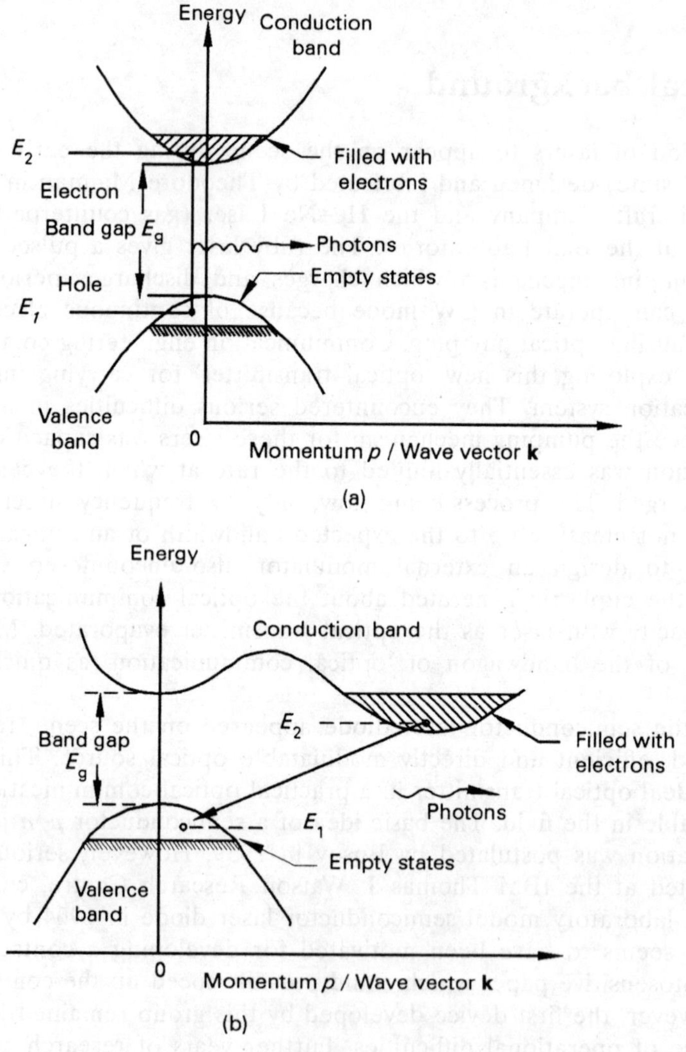

Fig. 3.1 *(a) Direct bandgap materials. (b) Indirect bandgap materials.*

the X-axis represents the crystal momentum p or the wave vector k for the electron in a crystal having a relationship

$$p = 2\pi hk \tag{3.1}$$

where h is the Planck's constant. However, the abscissa is often shown as the electron wavevector k rather than the momentum p. If and when the electron makes a transition downwards from the conduction band into the valence band and recombines with the hole, a photon is emitted with corresponding energy difference $E_2 - E_1$ given by

$$hf - E_2 - E_1 = E_g \tag{3.2}$$

However, in the case of indirect bandgap materials shown in Fig. 3.1(b), the lowest energy conduction band electrons have a non-zero momentum. Thus, the radiative recombination in this case requires momentum transfer due to photon participation. In such a case, the frequency of the emitted radiation is given by the equation

$$hf = E_2 - E_1; hf_p \approx E_g \tag{3.3}$$

where f_p is the photon frequency. Thus the recombination probability of an electron–hole pair leading to the emission of radiation is much smaller in indirect bandgap materials than in direct bandgap materials. Hence, we should use direct bandgap materials like gallium arsenide (GaAs) in the fabrication of semiconductor laser diodes.

3.3 The *p–n* junction

The commonly used semiconductor material for fabricating laser diodes is gallium arsenide which emits light at a wavelength (λ) of 8400 Å in the near infrared region of the optical spectrum. However, the wavelength of emission can be varied from 6300 Å (visible) for the III–V mixed semiconductor gallium arsenide phosphide to 8.5 μm (infrared) for the IV–VI semiconductor lead selenide.

In an intrinsic semiconductor material, the ambient temperature raises some electrons from the valence band into the conduction band by thermal excitation, leaving behind an equal number of holes in the valence band. However, the photon emission due to the recombination of electrons and holes (charge carriers) in intrinsic semiconductors is too small. One way to increase the photon emission is to increase the charge carrier density by adding dopants. This will lead to an energy-level diagram as shown in Fig. 3.2(a), resulting in an n-type semiconductor material when donor materials having excess electrons are added to it. On the other hand, when an accepter material rich in holes is added to it, a p-type semiconductor material is formed, as shown in Fig. 3.2(b). When these two materials are joined to form a p–n junction, the electrons and holes tend to move in opposite directions across the physical boundary by a diffusion process, setting up a depletion layer sandwiched between the p-type and n-type materials. Across this depletion layer, a barrier potential is set up in such a way that it inhibits further flow of charge carriers, thus establishing

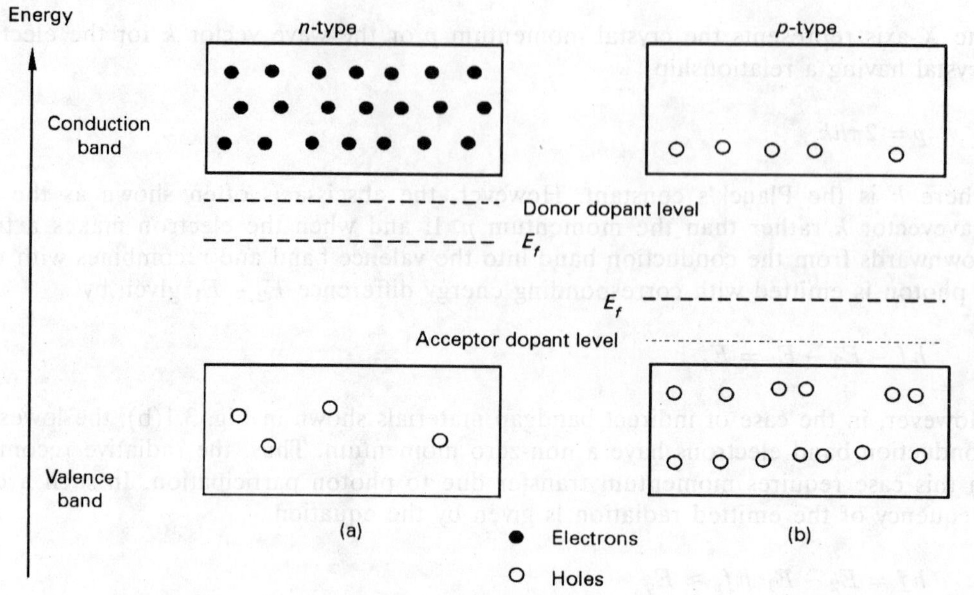

Fig. 3.2 Energy-level diagram for (a) n-type semiconductor material (b) p-type semiconductor material.

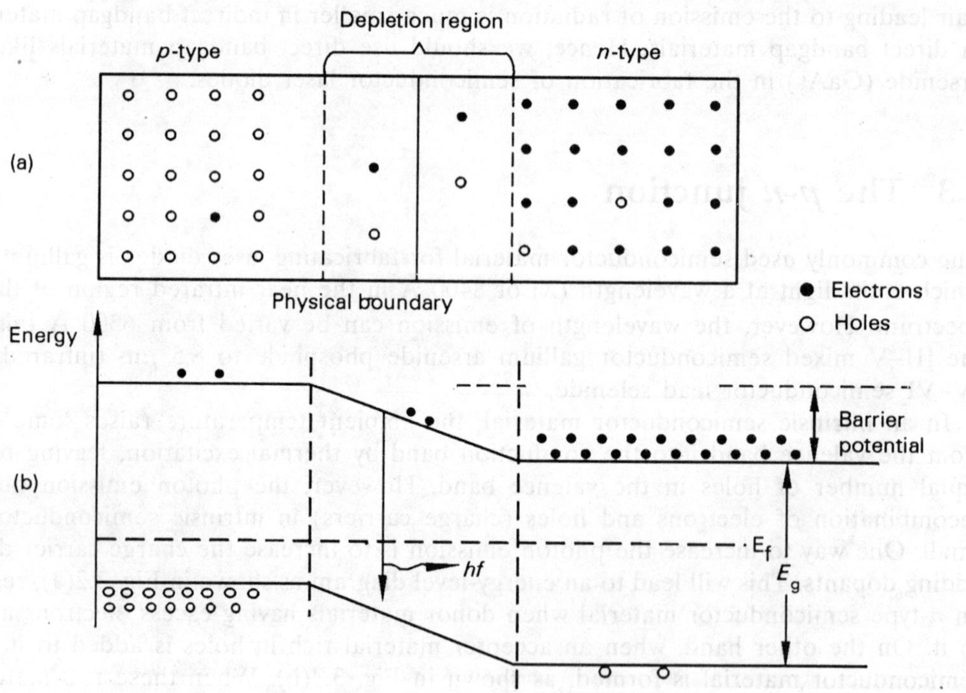

Fig. 3.3 (a) Typical distribution of charge carriers in a p–n junction (b) The corresponding energy-level diagram.

equilibrium condition. This situation is shown in Figs. 3.3(a) and (b). The electrons and holes which manage to find themselves in the depletion region may recombine with the emission of a photon corresponding to the energy gap E_g. In a normal p–n junction in electronic devices, these photons never see the light of the day since their numbers are very small and their efficiency of production is small in Si and Ge materials typically used in electronic devices.

3.4 Degenerate semiconductor diode

Now that we are interested in photon emission from the p–n junction, the obvious steps to be taken are the use of the direct bandgap semiconductor materials and increasing the supply of charge carriers by enhancing the doping levels. When heavy doping is done, the semiconductor material almost acts like a conductor and is thus referred to as degenerate semiconductor. So our photon device should use these 'degenerate' p-type and n-type diodes of bandgap semiconductor material whose energy-level diagram is shown in Fig. 3.4.

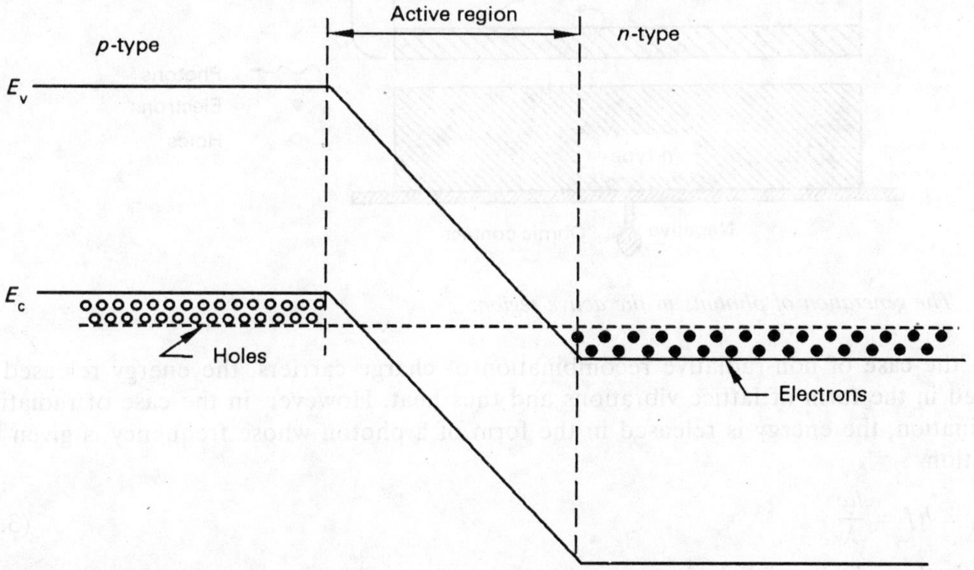

Fig. 3.4 Energy-level diagram of a p–n junction semiconductor diode with heavy doping (degenerate) without any external bias.

Just as in an ordinary semiconductor junction, the electrons and holes here also will tend to move towards the centre under the influence of diffusion, and will be balanced by the barrier potential set up across the junction. Since the charge carriers in this case are expected to recombine in the depletion region and generate photons, this region is now referred to as 'active region'. Now, when the p–n junction is forward biased, the charge carriers are injected into the active region from either side, producing photons as shown in Fig. 3.5. When the degenerate semiconductor p–n junction is forward biased, the holes are injected into the active region from the n-side. Thus these devices were referred to as 'injection laser diodes' or ILDs. Later, they were simply referred to as 'laser diodes' or

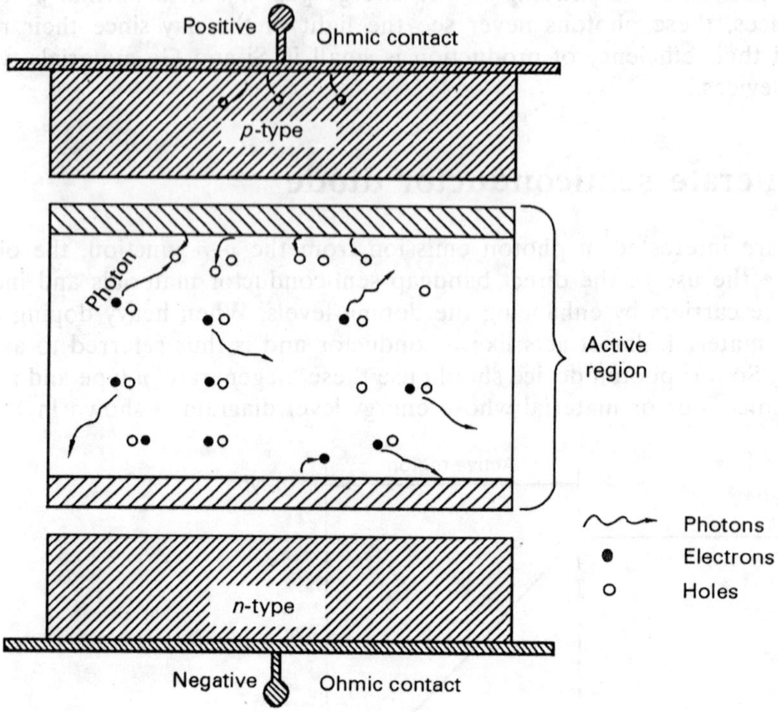

Fig. 3.5 The generation of photons in the active region.

LDs. In the case of non-radiative recombination of charge carriers, the energy released is dissipated in the form of lattice vibrations and thus heat. However, in the case of radiative recombination, the energy is released in the form of a photon whose frequency is given by the relation

$$E_g = hf = \frac{hc}{\lambda} \tag{3.4}$$

where c is the velocity of light in vacuum and λ is the optical wavelength. Substituting the approximate values of h and c in the above equation we get

$$\lambda = \frac{1.24}{E_g} \tag{3.5}$$

From this simple expression, the wavelength of the emission from different semiconductor materials can be quickly estimated. The units for λ and E_g are μm and eV, respectively. However, we should not forget that the photons emitted from the active region are due to the spontaneous emission and that they do not possess the coherence property though they are monochromatic. Besides, the active region is automatically formed around the physical boundary of the p–n junction whose dimensions are of the order of a few microns.

3.5 Estimation of forward bias voltage

When the degenerate semiconductor diode is forward biased, the potential 'hill' is decreased and the energy-level diagram will be as shown in Fig. 3.6. Under the influence of the external voltage, the Fermi level on the n-side is different from that of the p-side and it can be shown that

$$\phi_c - \phi_v = eV \tag{3.6}$$

where ϕ_c is the Fermi level in the conduction band of the n-side and ϕ_v is the Fermi level in the valence band of the p-side.

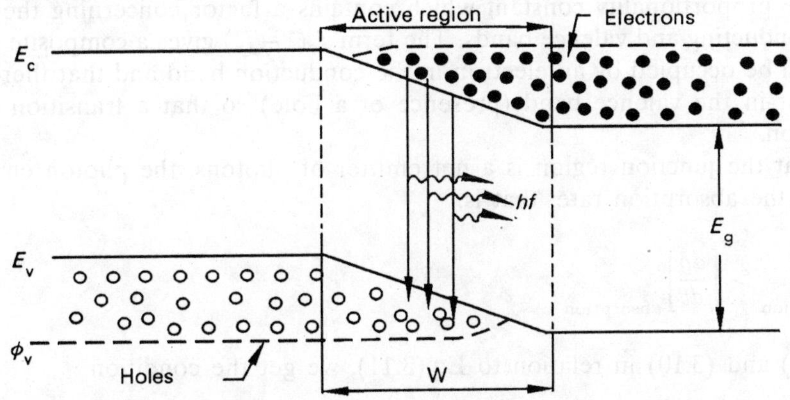

Fig. 3.6 Energy-level diagram of the forward biased degenerate semiconductor diode.

The distribution of charge carriers in semiconductor materials can be generally said to obey Fermi statistics. The probability that a state in the conduction band at energy E_c is occupied by an electron is given by the Fermi function:

$$f_c(E_c) = \left[1 + \exp\left(\frac{E_c - \phi_c}{kT}\right)\right]^{-1} \tag{3.7}$$

Similarly, the probability that an electron will occupy a state in the valence band on the p-side will be

$$f_v(E_v) = \left[1 + \exp\left(\frac{E_v - \phi_v}{kT}\right)\right]^{-1} \tag{3.8}$$

Assuming that the stimulated emission process is overwhelming, the contribution due to spontaneous emission can be ignored. If the stimulated emission coefficient B_{cv} is defined as the probability that a transition per unit time will occur from a state in the conduction band to one in the valence band, the inverse process, which is the stimulated absorption, will have a probability B_{vc}. The photon cloud density in the active region is $\rho(f)$, where f is frequency corresponding to the energy gap E_g.

Now the rate at which the photons are generated/absorbed due to the transition between the conduction band and valence band is given by

$$\left|\frac{d\rho}{dt}\right|_{\text{emission}} = CB_{cv} f_c(1 - f_v)\rho(f) \tag{3.9}$$

and

$$\left|\frac{d\rho}{dt}\right|_{\text{absorption}} = CB_{vc} f_v(1 - f_c)\rho(f) \tag{3.10}$$

where C is the proportionality constant which contains a factor concerning the density of states in the conducting and valence bands. The term $f_c(1-f_v)$ gives a composite probability that a state will be occupied by an electron in the conduction band and that there is a state which is vacant in the valence band (presence of a hole) so that a transition will occur, causing emission.

In order that the junction region is a net emitter of photons, the photon emission rate should exceed the absorption rate, that is,

$$\left[\frac{d\rho}{dt}\right]_{\text{emission}} > \left[\frac{d\rho}{dt}\right]_{\text{absorption}} \tag{3.11}$$

Using Eqs.(3.9) and (3.10) in relation to Eq.(3.11), we get the condition

$$CB_{cv} f_c(1 - f_v)\rho(f) > CB_{vc} f_v(1 - f_c)\rho(f) \tag{3.12}$$

Cancelling out the common terms on either side, this reduces to

$$f_c(1 - f_v) > f_v(1 - f_c) \tag{3.13}$$

which further leads to the following simple condition for net emission

$$f_c > f_v \tag{3.14}$$

Substituting the equations for the Fermi functions as given in Eqs.(3.7) and (3.8), we get

$$\left[1 + \exp\left(\frac{E_c - \phi_c}{kT}\right)\right]^{-1} > \left[1 + \exp\left(\frac{E_v - \phi_v}{kT}\right)\right]^{-1} \tag{3.15}$$

$$\text{or}\quad \exp\left[\frac{E_v - \phi_v}{kT}\right] > \exp\left[\frac{E_c - \phi_c}{kT}\right] \tag{3.16}$$

$$\text{or}\quad (\phi_c - \phi_v) > (E_c - E_v) \tag{3.17}$$

$$\text{Since } E_c - E_v = hf \tag{3.18}$$

$$\phi_c - \phi_v > hf \tag{3.19}$$

This implies that the difference in the Fermi level $(\phi_c - \phi_v)$ must exceed the photon energy in order to ensure that the active region acts like a net emitter of radiation. Recalling that the difference in the Fermi levels on either side of the junction is manifested due to the potential difference V across it, as given in Eq.(3.6), we can write

$$eV > hf \tag{3.20}$$

Example 3.1 For a GaAs sample emitting at 0.87 μm, calculate the bandgap energy and compute the minimum voltage that has to be applied across the junction in order to create a condition for net emission of photons from it.

Solution For a semiconductor device, it is shown that

$$E_g = hf = 6.6 \times 10^{-34}\text{J.s} \times \frac{3 \times 10^8\text{m/s}}{0.87 \times 10^{-6}\text{m}} = 2.28 \times 10^{-20}\ \text{J}$$

For the junction to act like a net emitter of photons, we have

$$V \geq \frac{hf}{e} = \frac{2.28 \times 10^{-20}\ \text{J}}{1.6 \times 10^{-19}\ \text{C}} = 0.142\ \text{eV}$$

3.6 Threshold current density

The photons that are generated in the active region lack direction because of spontaneous emission. To impart direction to the emitted light and, therefore to achieve laser output rather than just light emission, it is necessary to provide a semiconductor junction in the vicinity of the active region. This can be done by first cleaving the two sides of the crystal, making them parallel to each other, to a high degree of accuracy. Then by polishing these two faces of the crystal, the crystal–air boundary provides a natural reflecting surface. In order to channel the photons along the length of the device, the other faces along the breadth are roughly ground or deliberately etched or roughened, to eliminate reflections from these two sides. By further mirroring the back side of the device by coating it with a totally reflecting material, the laser beam can be made to emerge from the front surface of the device as shown in Fig. 3.7.

Because a large number of charge carriers move into the active region when the current flows across the junction due to external bias voltage, the net area of the active region is about 1 mm^2 and its width is typically 5 to 10 μm. These junction diode lasers are more efficient than the conventional lasers discussed in Chapter 2 because almost every hole–electron pair injected into the active region generates a useful photon. However, part of the electrical energy is lost in the form of heat across the electrical resistance internal and external to the diode. But not all the generated photons will contribute to the laser radiation emerging from the junction.

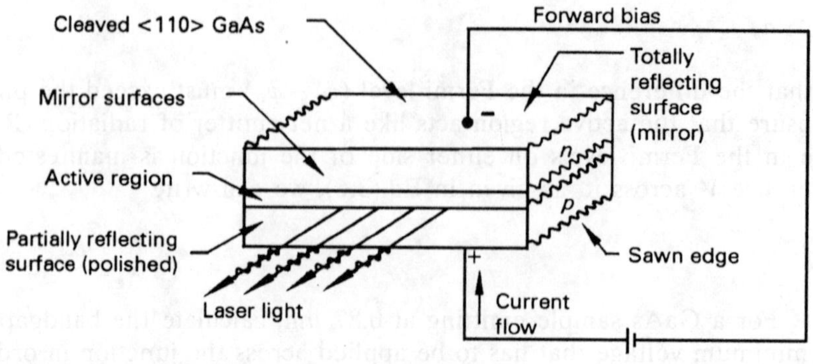

Fig. 3.7 Schematic diagram of a semiconductor laser diode.

In order to sustain laser action in the junction, the emitted photon rate must overcome the following losses :

(a) The absorption coefficient α_0 of the device for the photon generated.
(b) The losses due to transmission at the partially reflecting surface at the emitting end of the laser diode. [If R is the optical power reflection coefficient of this surface, then the transmission loss per unit length is given by $1/L\log(1/R)$, where L is the active layer length].
(c) The diffraction losses in the active region, which are approximately given by an empirical relation given below

$$\alpha_D = 0.42\left(\frac{\lambda}{nw^2}\right) \tag{3.21}$$

where n is the refractive index of the semiconductor material and w is the width of the active region.

If the total current passing through the active region from the positive terminal (p-side) to the negative terminal (n-side) when forward biased is I, then the current density J is expressed as

$$J = \frac{I \text{ (in amperes)}}{\text{area across which it is flowing in sq. metres}} \tag{3.22}$$

Thus the number of charge carriers injected into the active region per unit area per second is J/e. If the quantum efficiency η is the ratio of the number of photons produced per unit hole–electron pair, then the number of photons produced per unit area per second is given by $\eta J/e$. Hence, the number of photons produced in the active region per unit volume per second is $\eta J/ew$, where w is the width of the active region. If the effective line width of photon emission due to spontaneous emission is given as Δf, then the rate of spontaneous emission per unit frequency interval per unit volume is given by $J\eta/ew\Delta f$.

Since the active region of volume V (= area × w) acts like a resonator cavity, as per the theory of a resonator cavity, it can sustain a number of modes N_f per unit frequency interval per unit volume neglecting dispersion

$$N_f = \frac{8\pi\eta^3}{\lambda^2 c} \tag{3.23}$$

Thus the rate of spontaneous emission per mode is

$$\alpha = \frac{J\eta}{ew\Delta f N_f}$$

$$= \frac{J\eta}{ew\Delta f} \times \frac{\lambda^2 c}{8\pi\eta^3} \tag{3.24}$$

and hence the gain β per unit length of the active medium is

$$\beta = \frac{\eta\alpha}{c}$$

$$= \frac{J\eta\lambda^2}{8\pi\eta^2 ew\Delta f} \tag{3.25}$$

The point at which the gain in the medium is able to overcome the net loss in the medium is usually referred to as the threshold condition, which is represented in terms of the current density as J_{th}.

$$\frac{J_{th}\lambda^2}{8\pi\eta ew\Delta f} = \alpha_0 + \alpha_T + \alpha_{dif} \tag{3.26}$$

If the two ends of the optical resonator cavity on either side of the active region have reflection coefficients r_1 and r_2, then the transmission losses through them is given by the equation

$$\alpha_T = \frac{1}{L}\log\frac{1}{r_1 r_2} \tag{3.27}$$

$$\text{Therefore,} \quad J_{th} = \frac{8\pi\eta ew\Delta f}{\lambda^2}\left[\alpha_0 + \frac{1}{L}\log(1/r_1 r_2) + \alpha_{dif}\right] \tag{3.28}$$

However, if one end is mirrored with 100% reflecting material, and $r_1 = 1$ and $r_2 = R$, then

$$\alpha_T = \frac{1}{L}\log\frac{1}{R} \tag{3.29}$$

Since the partially reflecting side of the optical cavity is nothing but the semiconductor–air interface, the reflection coefficient can be calculated from the relationship

$$R = \frac{\sqrt{\epsilon_r} + 1}{\sqrt{\epsilon_r} - 1} \tag{3.30}$$

where ϵ_r is the ratio of the relative permittivity of the semiconductor material to the permittivity of free space.

Exercise Given the following data for the GaAs laser diode operating at room temperature, calculate: (i) diffraction losses in the cavity, (ii) attenuation due to the finite reflection of the end surfaces, and (iii) the threshold current density.

Width of the emitting region w	=	10 μm
Wavelength of emission	=	0.9 μm
Line width f	=	10^{13} Hz
Length of the cavity, L	=	0.1 mm
Quantum efficiency, η	=	30%
Dielectric constant of the material, ϵ_r	=	13
Breadth of the cavity, H	=	0.05 mm
α_0	=	20 nepers/metre

The first laser diode made was a *p–n* homojunction, meaning that the same intrinsic material is used on either side of the junction. The device was successfully operated under laboratory conditions and its output characteristics are shown in Fig. 3.8. However, heat was the major problem to contend with in the early devices. If unattended to, the increase in temperature could lead to an increasing demand in threshold current density, ultimately leading to device breakdown. The device had to be operated in a low temperature bath which made it unwieldy and cumbersome for field use. The overall power efficiency of the device is expressed as

$$\text{overall power efficiency} = \frac{\text{useful optical power output}}{\text{input electrical power}} \tag{3.31}$$

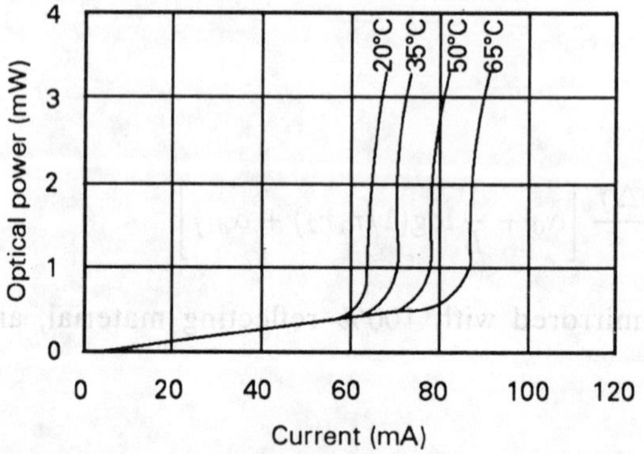

Fig. 3.8 Typical output characteristics of a laser diode.

The input electrical power is drawn from the normal input electrical connection. Thus this is referred to as 'plug-point' or 'plug-in' efficiency. Though the homojunction laser diode is more efficient than the conventional laser, it is still not preferred if operated in a CW mode because of the cooling requirements. Thus the IBM research group quickly lost interest in such a device for computer applications. However, it generated sufficient interest in the communication community to pursue the developments further. These difficulties were overcome in heterostructure devices as explained in Section 3.7.

3.7 Heterostructure devices

There are two important reasons for the efficiency of a homojunction laser diode being low:

(a) The photons generated in the active region of the p–n junction due to the recombination of charge carriers injected into it, may digress into the p and n regions of the devices and get absorbed, and not contribute to the laser output.
(b) The charge carriers injected from either side into the active region may not always recombine to yield a photon but instead may end up on either side of the fence without contributing to the laser output.

Thus, it stands to reason to expect that the efficiency of the device will improve considerably if these loopholes are plugged in. This is what is exactly attempted to be achieved in a heterostructure device. A heterostructure implies that more than one material is used in the fabrication of the device instead of one as in the case of the homostructure. The operation of the double heterostructure device in comparison with a homostructure device can be understood from Fig. 3.9. These figures are self-explanatory and clearly indicate how the charge carrier confinement is effected by creating a potential wall on either side of the active region. The photon confinement is due to the downward step in refractive index created on either side because of the enhanced bandgap of AlGaAs, which is basically GaAs.

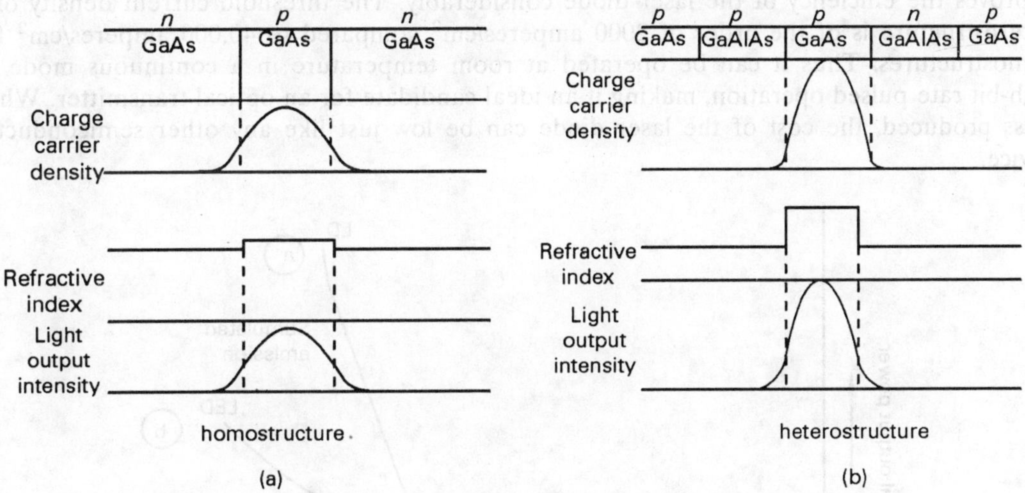

Fig. 3.9 Comparison between homostructure and heterostructure devices.

The selection of materials for making heterostructure devices has to be carefully done so as to ensure a close match between the crystal lattice parameters of the two adjoining heterostructures. Otherwise, the interfacial defects and strains in the device developed due to temperature changes might affect the radioactive efficiency and the lifetime of the laser diode. A heterojunction laser diode can be fabricated with ternary alloy GaAlAs by combining two semiconductor materials GaAs (E_g = 1.43 eV and a_0 = 5.64 Å) and AlAs (E_g = 2.16 eV and a_0 = 5.66 Å). The resulting material is $Ga_{1-x}Al_xAs$, where x represents the number of aluminium molecules which replace the gallium molecules. The energy gap of this alloy in electron volts can be given by the empirical relation

$$E_g = 1.424 + 1.266x + 0.266x^2 \tag{3.32}$$

where x can vary in the range $0 < x < 0.37$, if it were to remain as a direct gap material.

Example 3.3 Consider a $Ga_{1-x}Al_xAs$ heterostructure laser diode with $x = 0.07$. Calculate the energy gap.

Solution The energy gap E_g is given by

$$\begin{aligned} E_g &= 1.424 + 1.266x + 0.266x^2 \\ &= 1.424 + 1.266 \times 0.07 + 0.266 \times (0.07)^2 \\ &= 1.51 \text{ eV.} \end{aligned}$$

Thus, it is possible to tailor a device for optimal results. The double heterostructure improves the efficiency of the laser diode considerably. The threshold current density of a heterostructure is of the order of 2000 amperes/cm^2 compared to 40,000 amperes/cm^2 for homostructures. Thus it can be operated at room temperature in a continuous mode or high-bit rate pulsed operation, making it an ideal candidate for an optical transmitter. When mass produced, the cost of the laser diode can be low just like any other semiconductor device.

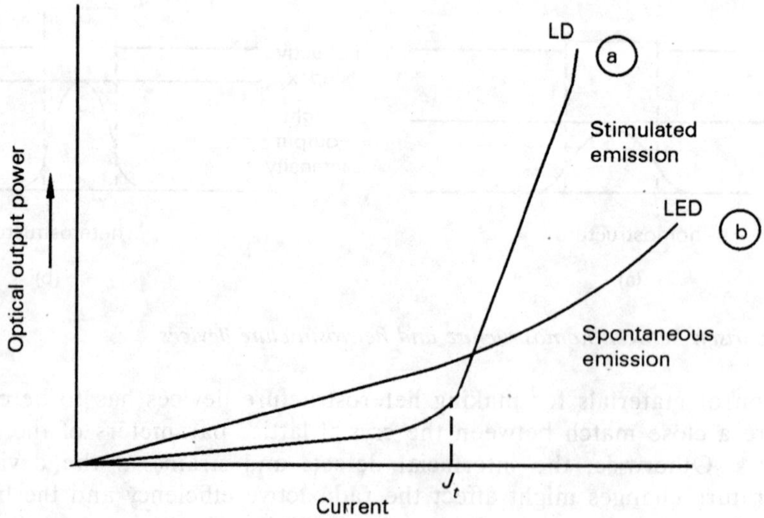

Fig. 3.10 Output characteristics of LD and LED.

The cost of the device is mainly contributed by the optical resonator cavity which has to be created across the active region by cleaving, mirroring and polishing. Thus, if the device is confined to just spontaneous emission criteria using the optical resonator cavity, the cost of the device can be drastically reduced. The penalty one pays for this is the reduced optical power output, which is not laser quality. Such a device is called Light Emitting Diode or LED. The output characteristics of LD and LED are given in Figs. 3.10(a) and (b). For transmission through low-loss medium using intensity modulation schemes, one might be able to get away with the use of LED as an optical transmitter. Besides lack of coherence,

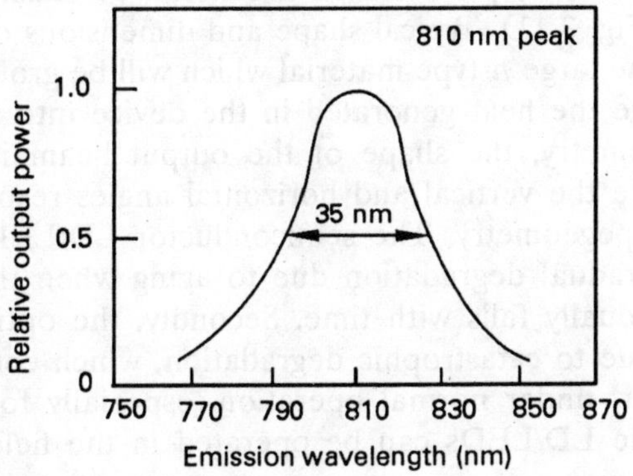

Fig. 3.11 Spectral characteristics of a typical LED.

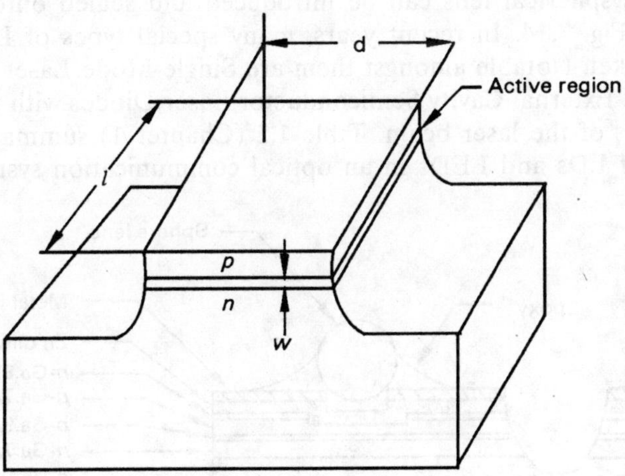

Fig. 3.12 Typical shape and dimensions of LD/LED.

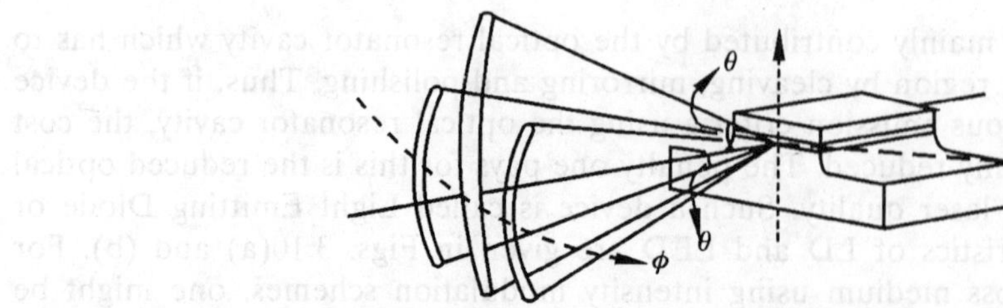

Fig. 3.13 Shape of the output optical beam from LD/LED.

the LED output is not spectrally pure though it can still be considered monochromatic by conventional standards (Fig. 3.11). Typical shape and dimensions of an LD/LED are shown in Fig. 3.12. Because of the large n-type material which will be ground when forward biased, it can effectively discharge the heat generated in the device into the ambient atmosphere. Because of the stripegeometry, the shape of the output beam is as shown in Fig. 3.13. The values of θ and ϕ are the vertical and horizontal angles respectively, of the radiation coming out from the stripegeometry. The semiconductor LD/LEDs suffer from two types of degradation. One is gradual degradation due to aging when the optical output for the same electrical input gradually falls with time. Secondly, the optical output suddenly and drastically comes down due to catastrophic degradation, which comes about by driving the device too long. However, under normal operation, especially for pulse mode operation, the commercially available LD/LEDs can be operated in the field for more than 100,000 hours without any difficulty.

Because of the non-uniform field power distribution in the output optical device, it is not always convenient to use LDs/LEDs in an optical communication system. The output can be improved by charging the stripegeometry of the active region by proper tailoring of the device. Otherwise, a spherical lens can be introduced and sealed onto a surface emitting device as shown in Fig. 3.14. In recent years, many special types of LDs and LEDs have appeared in the market. Notable amongst them are Single-Mode Laser Diodes, Distributed Feedback Lasers and External Cavity Semiconductor Laser Diodes with improvements in the output characteristics of the laser beam. Table 1.2 (Chapter 1) summarises the advantages and disadvantages of LDs and LEDs in an optical communication system.

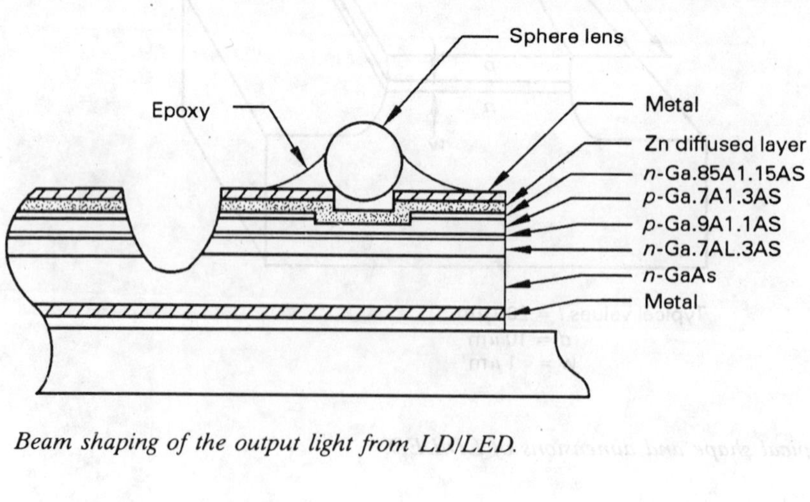

Fig. 3.14 Beam shaping of the output light from LD/LED.

4 Transmission medium: Atmospheric propagation

4.1 Historical background

The early models of optical communication were based on microwave communication systems with the atmosphere as transmission medium. Initially, it looked as though that it would not pose any difficulty because of the high directionality of the optical transmitter (laser) compared to those at lower frequencies. This means that the range of the optical communication link should be longer and the terminal equipment (antenna) be made smaller. Initial experiments that were carried out with atmospheric optical communication systems revealed that the atmosphere, for the propagation of optical laser beam through it, can create many complications. Basically, there should not be any physical obstacle present in the line of sight path between the transmitting and receiving antennae. Since the optical beam diameter is relatively small, any small obstacles like falling leaves, flying birds and airplanes can disrupt the optical communication link. As the frequencies involved are much higher, the energy of the photon, hf, is much larger than that of microwave frequencies; this energy is strong enough to cause electronic, atomic and molecular transitions in the atmospheric gases. As a result, the optical energy is selectively absorbed at certain frequencies corresponding to atmospheric energy bands. In addition to this, the photons can be scattered away from the transmission path since the wavelength of the photons has comparable dimensions with respect to the atoms/molecules of the atmospheric gases. Unguided optical communication system was the subject of a great deal of research and development activity in the sixties. In spite of the vagaries of the atmosphere, successful atmospheric optical communication links have been developed and installed and are still relevant for special applications.

4.2 Laser beam launching into the atmosphere

The laser beam can be launched into the atmosphere, like any other electromagnetic radiation, using an antenna. A cross-sectional view of a typical transmitting antenna is shown in Fig. 4.1.

The optical transmitting antenna has the following two important objectives besides transmitting/coupling optical energy from the laser into the transmission medium

(a) To act as a beam expander.
(b) To act as a beam collimator.

Since the laser beam diameter is very small (of the order of a few mm), it can easily miss the receiver due to disturbances in the atmosphere. It is therefore desirable to expand the beam sufficiently in spite of the corresponding decrease in optical energy density. Also, the laser beam has a natural divergence angle associated with it. It is, therefore, desirable to collimate the beam. Besides, the optical antenna acts like a front end to the optical transmitter, and the laser, which is relatively sensitive, is not directly exposed to the atmosphere.

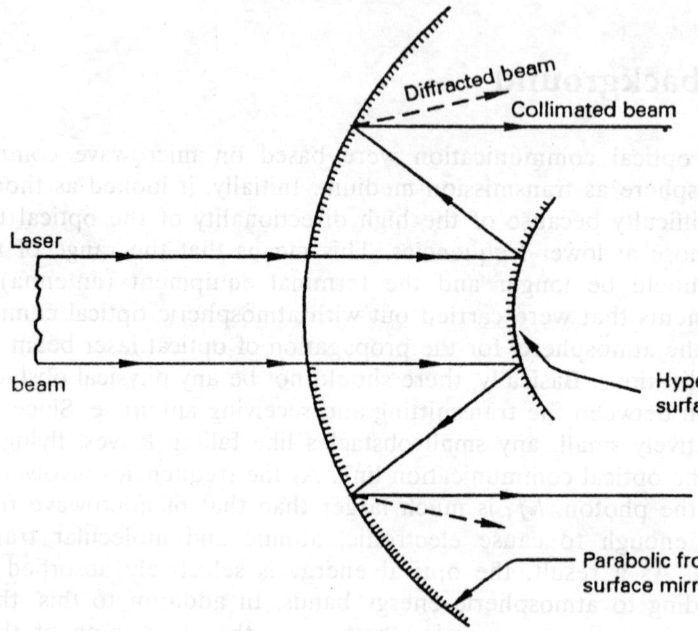

Fig. 4.1 Typical cassegranian type optical transmitting antenna with reflective optics.

Example 4.1 Assuming the electromagnetic radiation from a source as an inverse function of the square of frequency, compute the ratio of the spatial power densities on the receiver plane at optical frequencies (10^{14} Hz) and microwave frequencies (10^9 Hz). If the diameter of the optical antenna is 10 cm and the microwave antenna diameter is 10 m, calculate the ratio of power outputs from these two antennae.

Solution The divergence angle is given by (see Fig. 4.2)

$$\frac{\Omega_0}{\Omega_m} = \frac{f_m}{f_0}$$

$$\frac{P_0}{P_m} = \left(\frac{f_0}{f_m}\right)^2$$

$$= \left[\frac{10^{14}}{10^9}\right]^2 = 10^{10}$$

The received power output ratio is given by

$$\frac{P_{R0}}{P_{Rm}} = 10^{10} \times \frac{A_{R0}}{A_{Rm}} = 10^{10}\left[\frac{d_{R0}}{d_{Rm}}\right]^2$$

$$= 10^{10} \times \left[\frac{0.1}{10}\right]^2 = 10^6$$

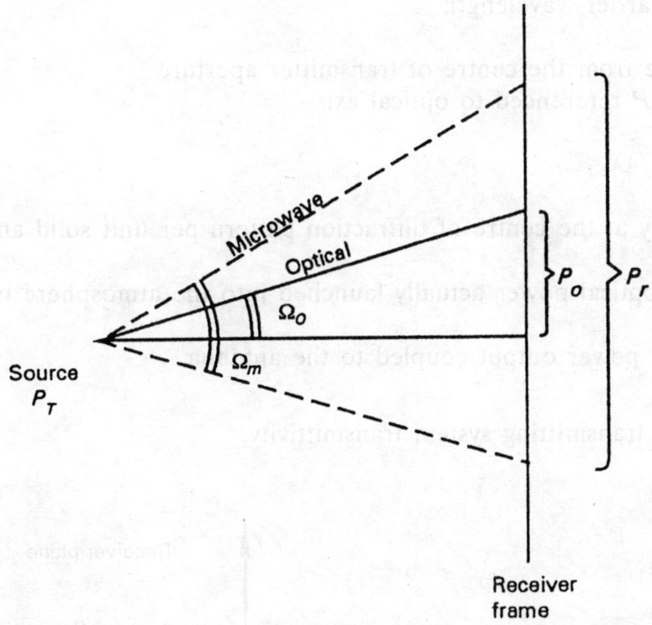

Fig. 4.2 The geometry of open air optical communication link.

This shows that the optical antenna, in spite of being 100 times smaller than the microwave antenna, gathers a million times more power. This means that for the same receiver sensitivity, the optical communication range can be much larger with much smaller terminal equipment.

When the range of the communication link (R) is large, i.e., in the far-field pattern, the diameter of the collimated beam is smaller in comparison with the diffracted beam pattern. Thus it is important to study the optical power distribution pattern on the receiver plane in order to design and position the receiver antenna for optimum performance of the optical communication link. For this analysis, the geometry assumed is as shown in Fig. 4.3.

The intensity per unit solid centred around the point P on the receiver plane as shown in Fig. 4.4 can be expressed in terms of the Bessel function as

$$I(P) = \left[\frac{2J_1(\chi)}{\chi}\right]^2 I(0) \tag{4.1}$$

where

$\chi = \pi d_T \alpha / \lambda_c$

$J_1(\chi)$ = first-order Bessel function of argument

d_T = transmitter aperture diameter

λ_c = optical carrier wavelength

α = half angle from the centre of transmitter aperture to point P referenced to optical axis

$$I(0) = \pi d_T^2 P_A / 4\lambda_c \tag{4.2}$$

= intensity at the centre of diffraction pattern per unit solid angle

where P_A = the optical power actually launched into the atmosphere by the antenna

P_L = laser power output coupled to the antenna

and $\tau_T = \dfrac{P_A}{P_L}$ = transmitting system transmittivity.

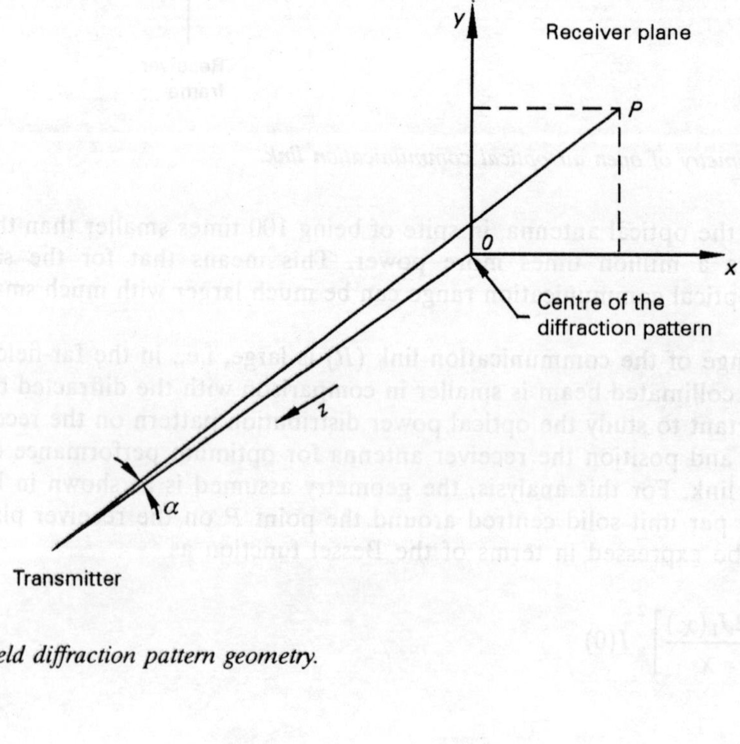

Fig. 4.3 Far field diffraction pattern geometry.

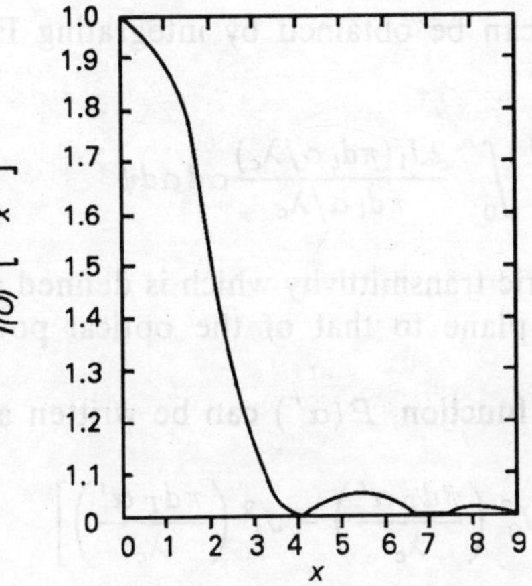

Fig. 4.4 The intensity distribution at any point P on the receiver plane in terms of the intensity at the centre.

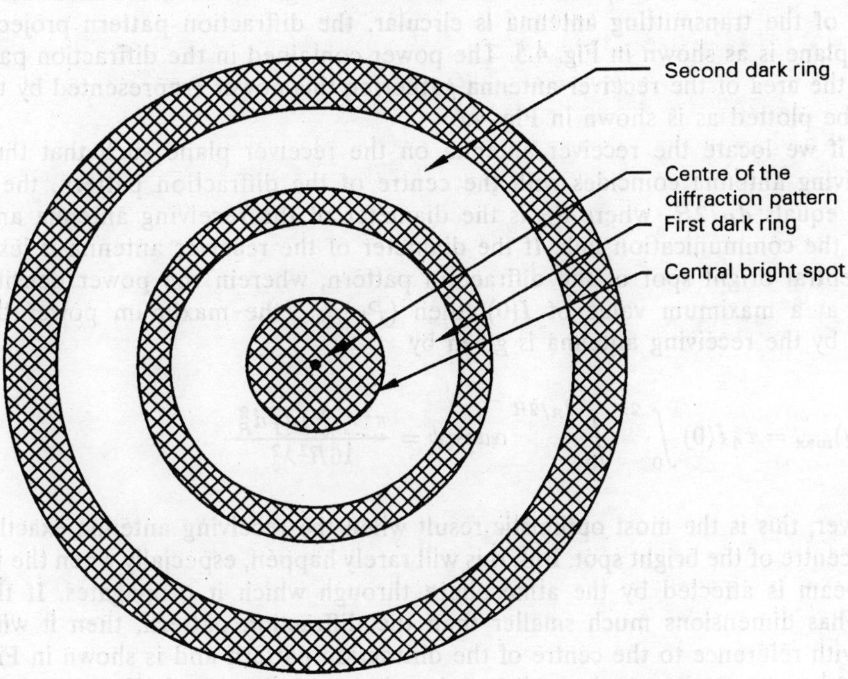

Fig. 4.5 Diffraction pattern for transmitting antenna with circular aperture

Figure 4.5 shows the diffraction pattern in terms of the intensity at any point P on the receiver plane with reference to the intensity at the centre of the diffraction pattern. The power incident on the receiver plane in an area which subtends an angle

α' as shown in Fig. 4.3 can be obtained by integrating Eq. (4.1) within the appropriate limits

$$P(\alpha') = \tau_A I(0) \int_0^{2\pi} \int_0^{\alpha} \frac{2J_1(\pi d_t \alpha/\lambda_c)}{\pi d_t \alpha/\lambda_c} \alpha d\alpha d\psi \tag{4.3}$$

where τ_A is the atmospheric transmittivity which is defined as the ratio of the optical power incident on the receiver plane to that of the optical power which left the transmitting antenna.

In terms of the Bessel function, $P(\alpha')$ can be written as

$$P(\alpha') = \tau_A P_A \left[1 - J_0^2\left(\frac{\pi d_T \alpha'}{\lambda_c}\right) - J_1^2\left(\frac{\pi d_T \alpha'}{\lambda_c}\right)\right] \tag{4.4}$$

The argument, $\pi d_T \alpha'/\lambda_c$ of the Bessel function can be denoted by χ'. Thus the term $[1 - J_0^2(\chi') - J_1^2(\chi')]$ represents a fraction of the total transmitted energy contained in a circle inside the diffraction pattern which subtends angle α' at the transmitter. Since the aperture of the transmitting antenna is circular, the diffraction pattern projected on the receiver plane is as shown in Fig. 4.5. The power contained in the diffraction pattern as we increase the area of the receiver antenna gathering this power (represented by the variable α') can be plotted as is shown in Fig. 4.6.

Now, if we locate the receiver antenna on the receiver plane such that the centre of the receiving antenna coincides with the centre of the diffraction pattern, the diffraction angle α' equals $d_R/2R$, where d_R is the diameter of the receiving antenna and R is the range of the communication link. If the diameter of the receiver antenna is less than that of the central bright spot of the diffraction pattern, wherein the power density is nearly constant at a maximum value of $I(0)$, then $(P_R)_{\max}$, the maximum power that can be gathered by the receiving antenna is given by

$$(P_R)_{\max} = \tau_A I(0) \int_0^{2\pi} \int_0^{d_R/2R} \alpha d\alpha d\psi = \frac{\pi^2 \tau_A P_A d_T^2 d_R^2}{16 R^2 \lambda_c^2} \tag{4.5}$$

However, this is the most optimistic result when the receiving antenna exactly coincides with the centre of the bright spot. But, this will rarely happen, especially when the transmitted optical beam is affected by the atmosphere through which it propagates. If the receiver antenna has dimensions much smaller than the diffraction pattern, then it will make an angle ϵ with reference to the centre of the diffraction pattern and is shown in Fig. 4.7. The illuminated area on the receiver plane taken between the two half-power points, where $I(P) = I(0)/2$, subtends an angle θ_T at the transmitter.

Letting $I(P) = I(0)/2$ in Eq.(4.1), the argument of the Bessel function can be obtained from the numerical techniques as

$$\frac{\pi d_T \theta_T}{2\lambda_c} \tag{4.6}$$

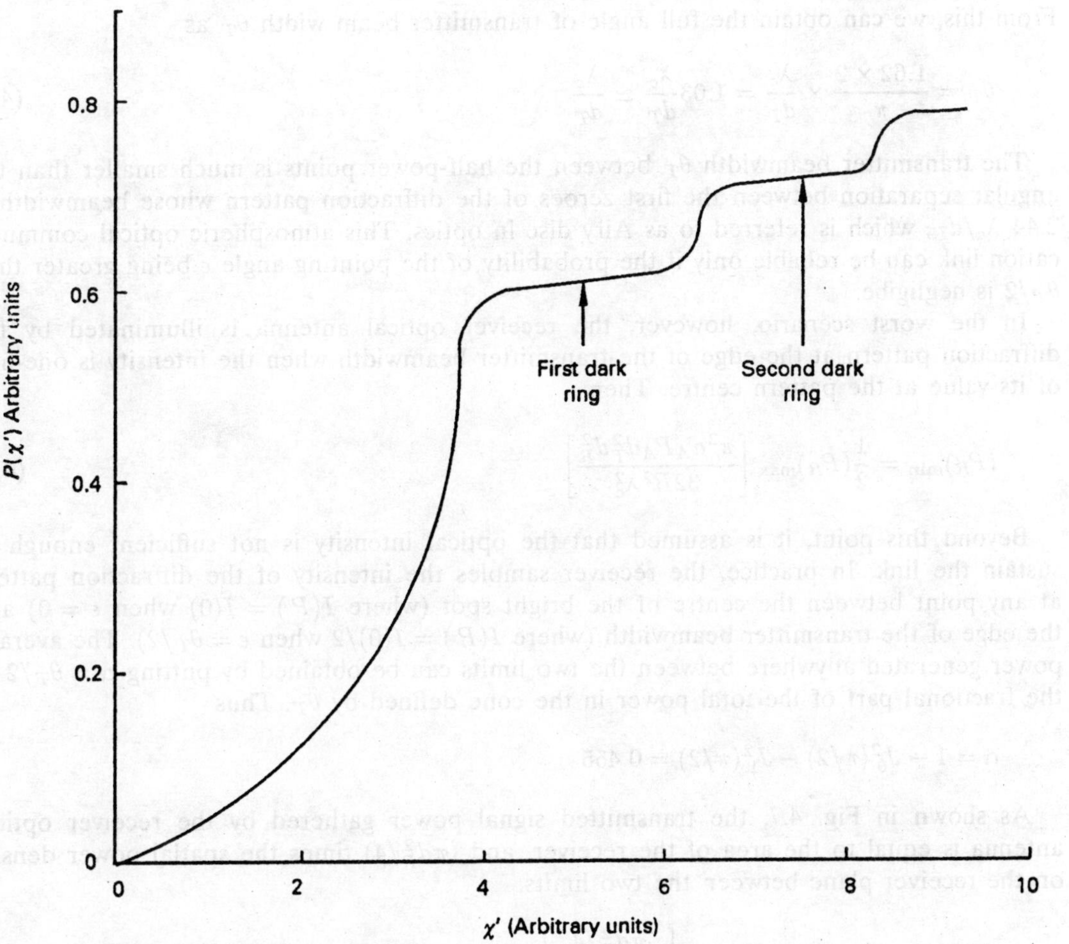

Fig. 4.6 ***The fraction of the laser power contained in a circle about the centre of the diffraction pattern.***

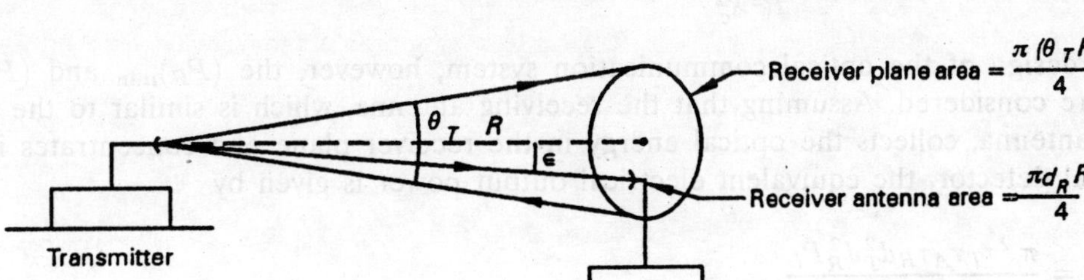

Fig. 4.7 ***The transmitted signal power gathered by the receiver optical antenna.***

From this, we can obtain the full angle of transmitter beam width θ_T as

$$\theta_T = \frac{1.62 \times 2}{\pi} \times \frac{\lambda_c}{d_T} = 1.03\frac{\lambda_c}{d_T} \simeq \frac{\lambda_c}{d_T} \tag{4.7}$$

The transmitter beamwidth θ_T between the half-power points is much smaller than the angular separation between the first zeroes of the diffraction pattern whose beamwidth is 2.44 λ_c/d_T, which is referred to as Airy disc in optics. This atmospheric optical communication link can be reliable only if the probability of the pointing angle ϵ being greater than $\theta_T/2$ is negligibe.

In the worst scenario, however, the receiver optical antenna is illuminated by the diffraction pattern at the edge of the transmitter beamwidth when the intensity is one-half of its value at the pattern centre. Then

$$(P_R)_{\min} = \frac{1}{2}(P_R)_{\max}\left[\frac{\pi^2 a_A P_A d_T^2 d_R^2}{32R^2\lambda_c^2}\right] \tag{4.8}$$

Beyond this point, it is assumed that the optical intensity is not sufficient enough to sustain the link. In practice, the receiver samples the intensity of the diffraction pattern at any point between the centre of the bright spot (where $I(P) = I(0)$ when $\epsilon = 0$) and the edge of the transmitter beamwidth (where $I(P) = I(0)/2$ when $\epsilon = \theta_T/2$). The average power generated anywhere between the two limits can be obtained by putting $\alpha = \theta_T/2$ as the fractional part of the total power in the cone defined by θ_T. Thus

$$\alpha = 1 - J_0^2(\pi/2) - J_1^2(\pi/2) = 0.455$$

As shown in Fig. 4.7, the transmitted signal power gathered by the receiver optical antenna is equal to the area of the receiver, and $(\pi d_R^2/4)$ times the spatial power density on the receiver plane between the two limits.

$$\begin{aligned}(P_R)_{\text{avg}} &= (0.455)\tau_A P_A\left[\frac{\pi d_R^2/4}{\pi\theta_T^2 R^2/4}\right]\\ &= (0.455)\frac{\tau_A P_A d_T^2 d_R^2}{R^2\lambda_c^2}\end{aligned} \tag{4.9}$$

For the design of the optical communication system, however, the $(P_R)_{\min}$ and $(P_R)_{\max}$ values are considered. Assuming that the receiving antenna, which is similar to the transmitting antenna, collects the optical energy in the receiver plane and concentrates it into an optical detector, the equivalent electrical output power is given by

$$P_D = \frac{\pi^2 \tau_T \tau_A \tau_R d_T^2 d_R^2 P_L}{32R^2\lambda_c^2} \tag{4.10}$$

where τ_R is the receiver system transmittivity which is the ratio of the optical power falling on the photodetector of the receiving antenna to the optical power falling on the receiving antenna.

4.3 Atmospheric attenuation

Atmospheric attenuation of the laser beam passing through it is described by the exponential law

$$a_A = \exp\{-\alpha_A L\} \tag{4.11}$$

where a_A is the atmospheric transmittivity, L is the atmospheric transmission path length and α_A is the attenuation coefficient which is also sometimes referred to as the extinction coefficient. The loss of laser radiation while propagating through the atmosphere can be due to absorption or scattering, and is given by

$$\alpha_A = \alpha_b + \alpha_s \tag{4.12}$$

where α_b = absorption coefficient and α_s = scattering coefficient. The atmospheric transmittivity a_A will then be a product of the absorption and scattering transmittivities.

$$a_A = a_b \times a_s \tag{4.13}$$

where a_b = absorption transmittivity and a_s = scattering transmittivity.

The absorption of laser radiation in the atmosphere is due to energy transitions in individual gaseous constituents like water vapour, carbon dioxide, ozone, etc. The atmospheric absorption of the optical radiation is selective in nature depending upon the absorption bands of these atmospheric gases and their relative concentration. Cross predictions of the atmospheric absorption can be made by estimating the relative concentrations of the molecular absorbants in the optical transmission path. But such estimates can only be approximate because the individual molecular concentrations in the atmosphere are a function of time (dynamic), depending upon the temperature and pressure. Besides, they also vary with the geographical location, altitude and prevailing weather conditions.

Atmospheric scattering of the optical radiation by gas molecules is a function of the wavelength of the optical radiation. When the wavelength of the optical radiation has dimensions approximately equal to that of gas molecules, it is referred to as Rayleigh's scattering. The attenuation coefficient due to Rayleigh's scattering α_{SR} is given by the empirical relation

$$\alpha_{SR} = 0.827 N A_p^3 \lambda_c^{-4} \tag{4.14}$$

where N is the number of particles per unit volume in the optical path and A_p is the cross-sectional area of the scattering particle.

Here, the particle area (A_p) and the wavelength (λ_c) are given in centimetres. As the Rayleigh scattering coefficient given in Eq.(4.14) is inversely proportional to λ_c^4, it implies that shorter wavelength radiation is more scattered or affected than longer wavelengths.

When the scattering is due to particles such as smoke and fog whose particle size is large compared to the transmission wavelengths, it is referred to as Mie scattering. Mie scattering is described by the following empirical relation

$$\alpha_{SM} = \frac{3.91}{V}\left[\frac{\lambda_c}{0.55}\right]^{-0.585^{-1/3}} \tag{4.15}$$

where V is the visual range in kilometres, α_{SM} is the Mie scattering coefficient in kilometres, and λ_c is the carrier wavelength in micrometres. The visual range measurements can be obtained from the local weather bureau.

Example 4.2 For a 1 km long optical communication link through the atmosphere, where the visual range is given as 5 km, estimate the Mie scattering coefficient and scattering transmittivity as a function of wavelength from 0.5 to 10 μm in finite steps.

Solution For estimating the Mie scattering coefficient, α_{SM} can be estimated using the empirical relation

$$\alpha_{SM} = \frac{3.91}{V}\left[\frac{\lambda_c}{0.55}\right]^{-0.585^{-1/3}}$$

where $V = 5$ km, and the equation for the scattering transmittivity

$$a_{SM} = \exp(-\alpha_{SM}L)$$

where $L = 1$ km. Taking the values of α_{SM} to be 0.5, 1.0, 2.0, 5.0 and 10.0 microns, it can be seen that the Mie scattering coefficient decreases with increasing wavelength. Therefore, the corresponding transmittivity increases.

Figure 4.8 is a plot of the per cent transmission of optical radiation through a typical atmosphere as a function of wavelength, considering that both absorption and scattering

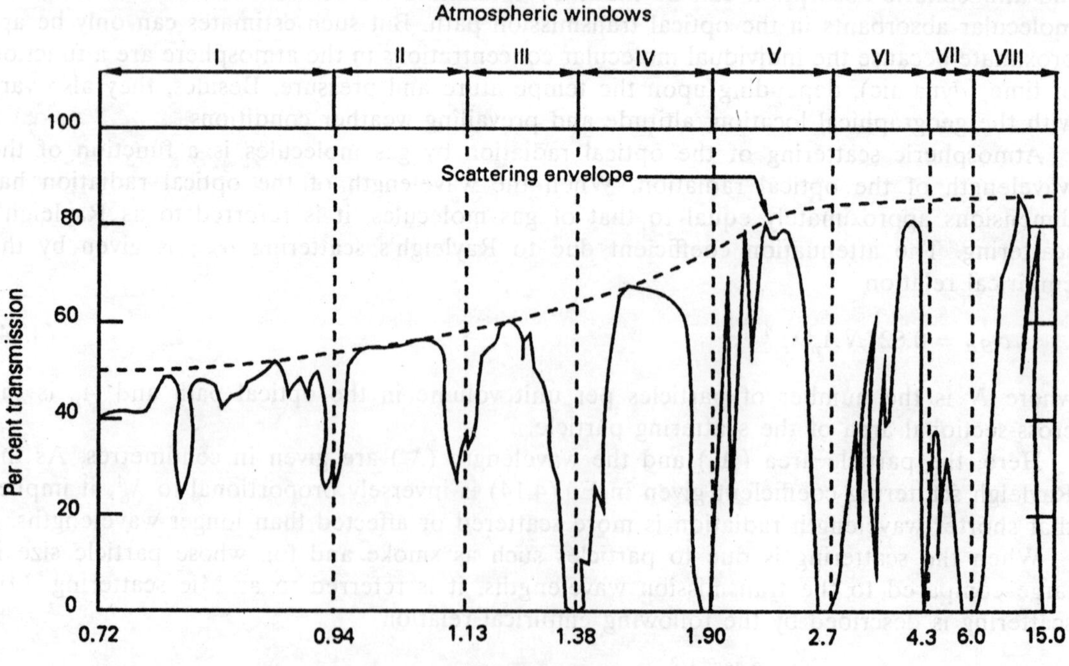

Fig. 4.8 Per cent transmission of the optical beam through the atmosphere as a function of wavelength.

selective absorptions by the atmospheric gases account for the sharp dips in the transmittivity curve. However, the gradual increase in the transmittivity with increasing wavelength is attributed to the scattering process discussed above. Figure 4.8 can be broadly divided into eight regions of relatively high transmittivity which are commonly referred to as atmospheric windows. While attempting to set up an optical atmospheric communication link, the designer is advised to refer to the figure for selecting a suitable laser as an optical transmitter.

4.4 Atmospheric turbulence

Part of the sunlight incident upon the earth is absorbed by the atmosphere. This absorption causes layers of the atmosphere to get heated up, causing convection currents, leading to turbulence. This turbulence is caused by temperature fluctuations (between different regions of the atmosphere) and the presence of the winds. All these factors cause the refractive index to vary randomly with respect to both space and time.

As the refractive index varies randomly from point to point, a laser beam propagating through such a medium will be deflected from its original path owing to a number of random reflections and refractions. The interaction of the laser beam with the atmosphere leads to random amplitude and phase variation of the laser carrier. The atmospheric turbulence on an optical communication link causes the following:

(a) *Beam steering*: Angular deviation of the beam from the line of sight path causing the beam to miss the receiver.

(b) *Image dancing*: Variations in the beam arrival angle causing the focus point to move in the image plane.

(c) *Beam spreading*: Small-scale scattering, increasing the beam divergence and causing a decrease in the spatial power density of the receiver.

(d) *Beam scintillation*: Small destructive interference within the beam cross-section, causing variations in the spatial power density at the receiver.

(e) *Spatial coherence degradation*: Losses in phase coherence across the beam phase fronts, degrading the photomixing performance.

As the time taken by the laser beam to propagate through the channel is infinitesimally small, the temporal variations in the refractive index can be ignored for this small time duration, and it will be sufficient to consider only the spatial variations of the index of refraction.

The relationship between the index of refraction of the atmosphere and the temperature is given by

$$n - 1 = 77.6\frac{P}{T}\,[1 + 7.52 \times 10^3\lambda^2] \times 10^{-6} \tag{4.16}$$

where P = pressure in millibars

T = temperature in degrees Kelvin

λ = light wavelength in microns.

For engineering applications, however, the following relation holds:

$$n - 1 = 7.8 \times 10^{-5} P/T \tag{4.17}$$

Random fluctuations in the atmospheric refractive index are induced by the corresponding variations in the atmospheric temperature, resulting in the naturally occurring random fluctuations in wind velocity, called turbulence. We can, therefore, express the index of refraction $n(r)$ as the sum of its free space value plus a random fluctuating component due to the turbulence:

$$n(r) = n_0 + n_1(r) \tag{4.18}$$

where $n_1(r)$ is the fluctuating component and is a function of the position r within the channel.

Because the turbulence is a random process, it must be described in terms of the statistical quantities, particularly, the structure function and the spatial correlation function. The degree of atmospheric turbulence and its relationship to the optical properties can be characterized by Tatarski's structure constant for refractive index fluctuations $C_n(r)$. The value of the above structure constant varies with the altitude and time of the day. Typical values for daytime conditions near the earth's surface are as given below:

Weak turbulence $C_n = 8 \times 10^{-9}\ \mathrm{m}^{-1/3}$

Intermediate turbulence $C_n = 4 \times 10^{-8}\ \mathrm{m}^{-1/3}$

Strong turbulence $C_n = 5 \times 10^{-7}\ \mathrm{m}^{-1/3}$

The greater the value of C_n, the greater is the degradation in the performance of the optical communication link due to atmospheric turbulence.

4.5 Parameters of atmospheric propagation

Many parameters are necessary to derive the propagation of a light beam or a light pulse through a scattering medium. If a collimated light beam is incident on a scattering medium, then after passing through the medium, the light will have the following problems

(a) *Spatial spreading*: The light beam will be physically widened.

(b) *Angular spreading*: The light beam will be decollimated.

(c) *Attenuation*: The light beam will be less intense.

Spatial spreading and angular spreading are interrelated in that angular spreading inside the scattering medium produces spatial spreading as the light propagates. Attenuation occurs because the light is scattered out of the incident beam.

When a laser beam passes through a medium containing a random distribution of discrete scattering centres, its initial angular profile will get spread. If the radiation is time dependent, as in a pulsed laser, multiple scattering will have an important effect. Since the scattered radiation travels a longer path on its way to the receiver than the unscattered light of the discrete beam, it experiences an additional delay, thereby distorting the time dependence of the laser at the receiver. Different distances along the various possible scatter paths imply different transit times. Therefore, light from a transmitted pulse will leave the scattering medium over the source range at time intervals longer than the incident pulse. This variation in transit times effectively broadens the incident pulse, producing an effect called multipath time spreading due to scattering and turbulence in a very critical manner and places an upper limit on pulse rate resolution. This, in turn, means that the channel capacity of an optical communication link is drastically curtailed by the atmosphere.

After frustrating experimentation with the atmospheric optical communication links during the early 1960s, it was soon realised that only guided wave propagation can realise the full potential of an optical communication link. This ultimately led to the development of *fibre-optic communication link*.

5 Transmission medium: Fiber optics

5.1 Historical background

Guided wave propagation of communication systems is nothing new at the lower end of the electromagnetic spectrum. Even today, most electronic communication is carried by copper cables whether twisted pairs, coaxial cables or copper waveguides. The idea of guiding optical beams came interestingly enough from a civil engineer in the 19th century! Figure 5.1 shows the idea of Norman French, an American civil engineer, to light up all the colonial houses of America using an arc lamp located in the cellar, and guiding the light from it to various rooms of the house by internally mirrored tubes. The idea of French was even patented and was considered as a big improvement over the conventional way of lighting up the buildings by gas tubings or candle lights. Later, the idea of using guided-wave propagation for optical beams came from research groups at the Bell laboratories. In order to keep the vagaries of the atmosphere at bay, they devised a long tube connecting the transmitter and receiver with convex lenses embedded at periodic intervals as shown in Fig. 5.2. This arrangement would take care of the natural divergence of optical beams by converging them at regular intervals. This type of optical 'pipe' was experimented with an optical waveguide in the mid-sixties as an alternative transmission medium to the atmosphere. Though, in principle, it was functioning well, its attenuation per kilometre was of the order of a few thousands of dB/km which is too high for practical applications. Further, it is too expensive, delicate and unwieldy for field use. However, the important breakthrough in guided-wave propagation came from a suggestion by Kao and Hockam in 1966 that a dielectric waveguide can act as a medium of transmission. This dielectric medium was soon referred to as fiber optics made out of glass which is highly transparent to an optical beam. It has been claimed that the pattern of independent multiple discoveries in sciences is dominant rather than a subsidiary one. The development of the semiconductor laser and fiber optics in the 1960s is a typical case to this effect. Together, they have not only revived optical communication but have also revolutionalized the field of communications.

5.2 Optical fiber attenuation

The extremely low attenuation or transmission loss of optical fibers is one of the most important factors in bringing about its wide acceptance as a medium of transmission. Signal attenuation within optical fibers, as with metallic conductors, is usually abbreviated as dB. The decibel (dB) is a convenient way of comparing two divergent power levels, say, P_1 & P_2. This is defined as

$$\text{attenuation in dB} = 10 \log_{10} \frac{P_2}{P_1} \tag{5.1}$$

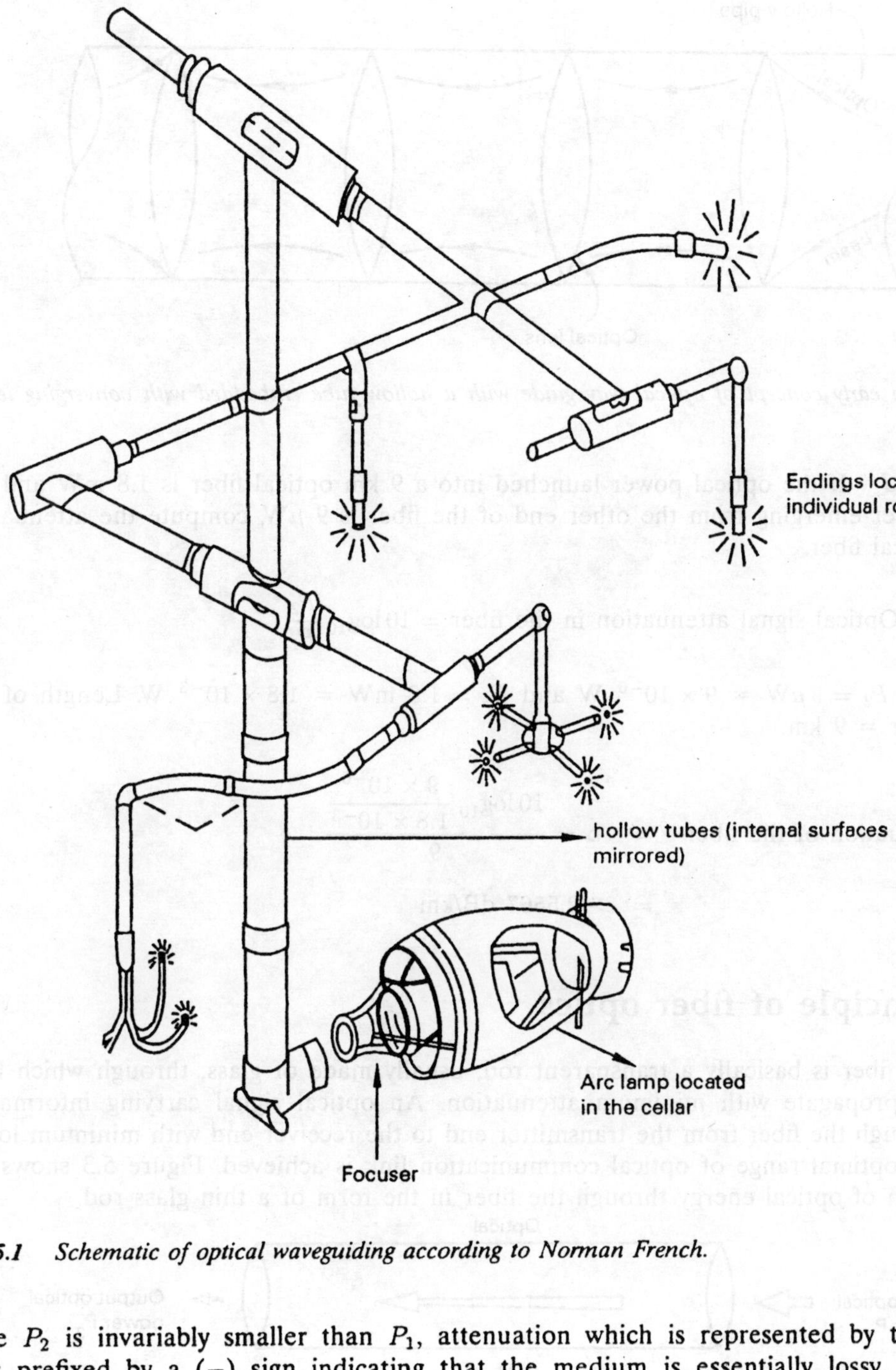

***Fig. 5.1** Schematic of optical waveguiding according to Norman French.*

Since P_2 is invariably smaller than P_1, attenuation which is represented by the symbol α, is prefixed by a (–) sign indicating that the medium is essentially lossy. In optical communication, the attenuation is usually expressed in decibels per unit length (i.e dB/km or dB km^{-1})

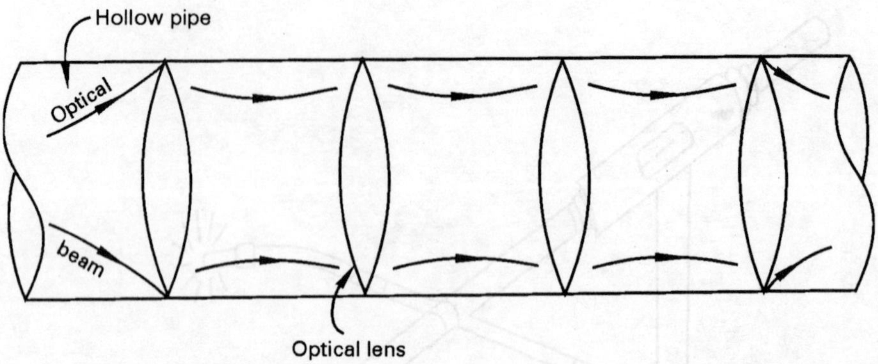

Fig. 5.2 *An early concept of optical waveguide with a hollow tube embedded with converging lenses periodically.*

Example 5.1 If the optical power launched into a 9 km optical fiber is 1.8 mW and the optical power emerging from the other end of the fiber is 9 μW, compute the attenuation of the optical fiber.

Solution Optical signal attenuation in the fiber $= 10\log_{10}\frac{P_2}{P_1}$

Here, $P_2 = 9\,\mu\text{W} = 9\times10^{-6}$ W and $P_1 = 1.8$ mW $= 1.8\times10^{-3}$ W. Length of the optical fiber = 9 km.

$$\text{Attenuation of the fiber} = \alpha = \frac{10\log_{10}\dfrac{9\times10^{-6}}{1.8\times10^{-3}}}{9}$$

$$= -2.5567 \text{ dB/km}$$

5.3 Principle of fiber optics

An optical fiber is basically a transparent rod, usually made of glass, through which light can easily propagate with minimum attenuation. An optical signal carrying information travels through the fiber from the transmitter end to the receiver end with minimum losses so that an optimal range of optical communication link is achieved. Figure 5.3 shows the transmission of optical energy through the fiber in the form of a thin glass rod.

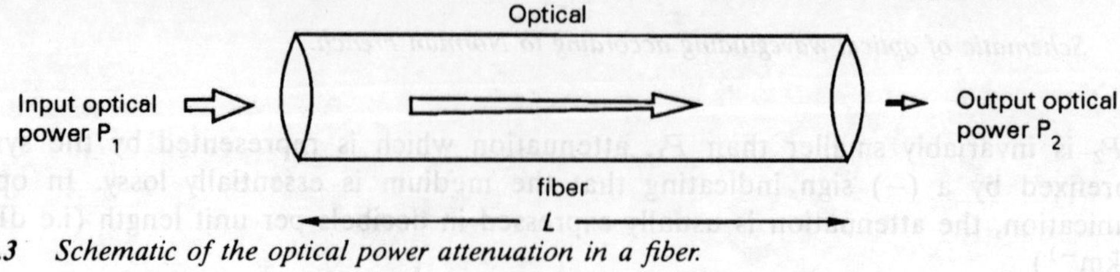

Fig. 5.3 *Schematic of the optical power attenuation in a fiber.*

In an unclad fiber, only a small part of the light is confined to the fiber, with most of the light leaking to the surrounding [Fig. 5.4(a)]. Thus the achievable optical communication range is rather limited. However, by providing a 'cladding' surrounding the core as shown in Fig. 5.4(b), the optical energy can be made to travel towards the receiver more effectively, thereby achieving a larger optical range.

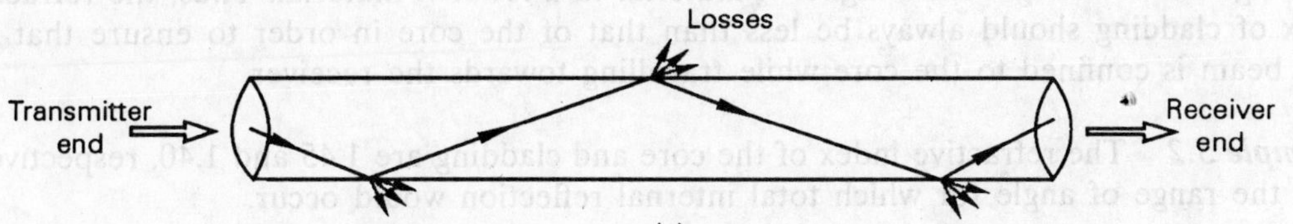

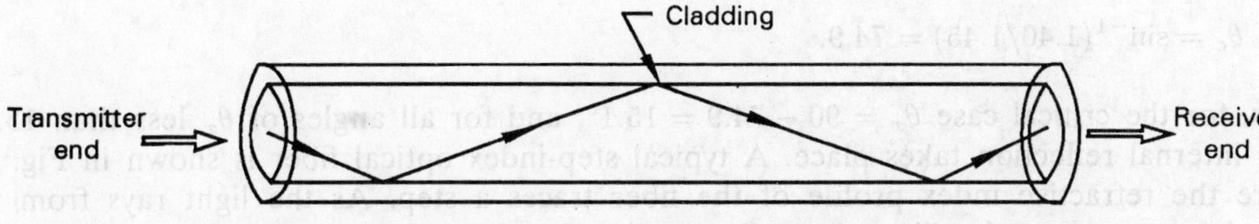

Fig. 5.4 Light waveguides (a) Simple fiber (b) Fiber with cladding.

When a ray travels across a boundary between two materials with different refractive indices n_1 and n_2, both refraction and reflection take place. In the case where $n_1 > n_2$, when light travels from a high to a low refractive index material, then according to Snell's law, the relation between θ_1 and θ_2 is given by

$$n_1 \sin\theta_1 = n_2 \sin\theta_2 \tag{5.3}$$

or
$$\frac{\sin\theta_1}{\sin\theta_2} = \frac{n_2}{n_1} \tag{5.4}$$

As θ_1 increases, a situation will arise when for a particular angle of θ_1, $\theta_2 = 90°$, the incident beam is entirely confined to the n_1 material without entering the n_2 material at all. The incident angle at which the above condition is satisfied is referred to as the critical angle θ_c.

$$\theta_1 = \theta_c, \quad \text{when} \quad \theta_2 = 90° \tag{5.5}$$

Equation (5.4) will then become

$$\sin\theta_c = \frac{n_2}{n_1}$$

or $\theta_c = \sin^{-1}\frac{n_2}{n_1}$ (5.6)

Since n_2 is less than n_1, the fraction (n_2/n_1) is less than 1 and θ_c will be less than 90°. When $\theta_1 > \theta_c$, the incident light is reflected and the light beam is entirely confined to medium 1. This condition is referred to as 'total internal reflection' and can occur only when light is travelling from a higher n material to a lower n material. Thus, the refractive index of cladding should always be less than that of the core in order to ensure that the light beam is confined to the core while travelling towards the receiver.

Example 5.2 The refractive index of the core and cladding are 1.45 and 1.40, respectively. Find the range of angle for which total internal reflection would occur.

Solution The critical angle θ_c is given by

$$\theta_c = \sin^{-1}(1.40/1.45) = 74.9.$$

Thus, for the critical case $\theta_x = 90 - 74.9 = 15.1°$, and for all angles of θ_x less than 15.1°, total internal reflection takes place. A typical step-index optical fiber is shown in Fig. 5.5 where the refractive index profile of the fiber traces a step. As the light rays from the optical source enter the fiber through air, they first bend inwards (towards the normal) as per Snell's law and then get reflected at the core–cladding boundary. As the incident angle θ_x increases, there reaches a point when the limit of total internal reflection reaches beyond which the light rays simply get refracted into the cladding.

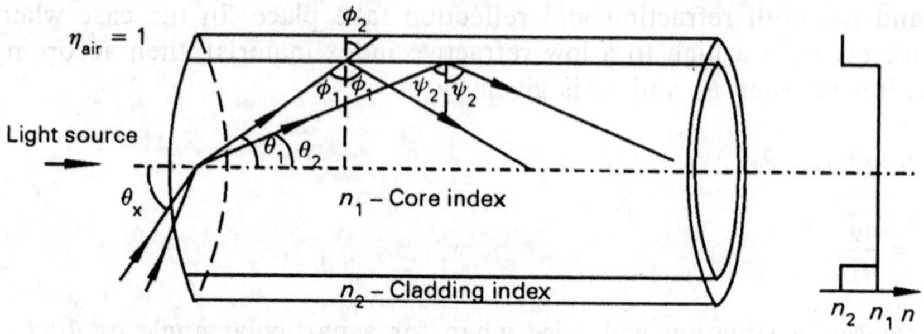

Fig. 5.5 Propagation of light through step-index fiber.

$n_0 \sin\theta_x = n_1 \sin\theta$, at the air–core interface, where $n_0 = 1$ for air.

$n_1 \sin\phi_1 = n_2 \sin\phi_2$, at core–cladding interface, where ϕ_2 is the angle of refraction in the cladding.

At critical angle, $\phi_2 = 90°$. Hence

$$n_1 \sin\phi_1 = n_2$$

From the geometry in Fig. 5.5, we have

$$\sin\phi_1 = \sin(90 - \theta_1) = \cos\theta_1 = (1 - \sin^2\theta_1)^{1/2}$$

Therefore, $n_2/n_1 = (1 - \sin^2\theta_1)^{1/2}$

At critical angle when $\theta_x = \theta_c \quad \theta = \theta_1$

Thus $\sin\theta_c = n_1 \sin\theta_1$

and $\dfrac{n_2}{n_1} = \left[1 - \dfrac{\sin^2\theta_c}{n_1^2}\right]^{1/2}$

or $\sin^2\theta_c = n_1^2\{1 - (n_2/n_1)\}$

$$\sin\theta_c = (n_1^2 - n_2^2)^{1/2}. \tag{5.7}$$

In the traditional optical sense, the numerical aperture of the waveguide is defined as

$$NA = \sin\theta_c = \{n_1^2 - n_2^2\} \tag{5.8}$$

However, if we take a three-dimensional view, only rays within a cone get coupled into the core of the optical fiber and travel towards the receiver. Such a cone is referred to as the acceptance cone and the angle of incidence of this cone into the fiber is referred to as the acceptance angle (Fig. 5.6).

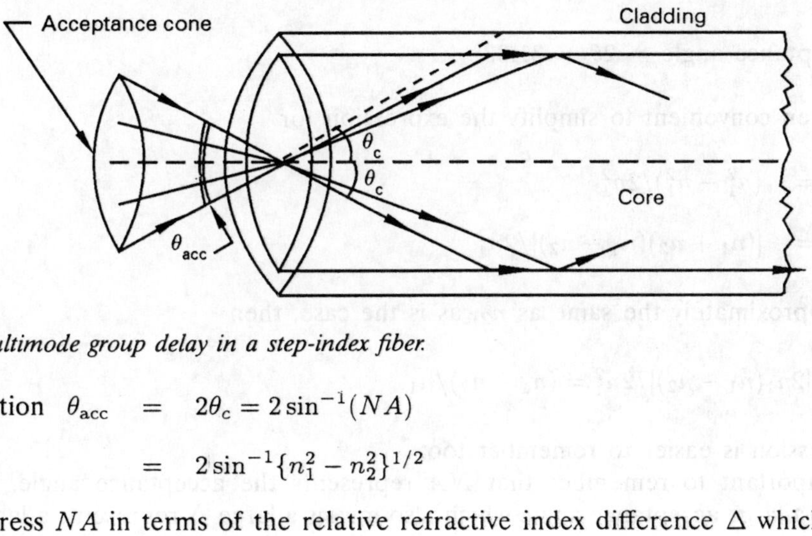

Fig. 5.6 Multimode group delay in a step-index fiber.

By definition $\theta_{acc} = 2\theta_c = 2\sin^{-1}(NA)$

$$= 2\sin^{-1}\{n_1^2 - n_2^2\}^{1/2}$$

We can express NA in terms of the relative refractive index difference Δ which is defined as

$$\Delta = (n_1^2 - n_2^2)/2n_1^2 = (NA)^2/2n_1^2$$

or $(NA)^2 = 2n_1^2\Delta$

Thus $NA = n_1(2\Delta)^{1/2}$. (5.9)

Example 5.3 Find the critical angle for an optical fiber with core refractive index of 1.43 and cladding refractive index of 1.40. Compute the numerical aperture (NA) and the acceptance angle for the same.

Solution $\sin\theta_c = \{n_1^2 - n_2^2\}^{1/2}$

$$\theta_c = \sin^{-1}\{n_1^2 - n_2^2\}^{1/2}$$

$$= \sin^{-1}\{(1.43)^2 - (1.40)^2\}^{1/2}$$

$$= 16.9°.$$

Acceptance angle $= 2\theta_c = 33.88°$.

Numerical aperture $= NA = \sin\theta_c = 0.2907$.

Example 5.4 A fiber has the following characteristics: $n_1 = 1.35$ (core index) and $\Delta = 2\%$ (or, a decimal value 0.02). Find NA and the acceptance angle.

Solution $NA = n_1 \times (2\Delta)^{1/2} = 1.35 \times (2 \times 0.02)^{1/2} = 0.27$

$$\theta_c = \sin^{-1}(NA) = \sin^{-1}(0.27) = 15.66°$$

Acceptance angle $= 2\theta_c = 31.33°$

It is often convenient to simplify the expression for

$$\Delta = (n_1^2 - n_1^2)/2n_1^2$$

$$= [(n_1 + n_2)(n_1 - n_2)]/2n_1^2$$

If n_1 is approximately the same as n_2, as is the case, then

$$\Delta \approx [2n_1(n_1 - n_2)]/2n_1^2 = (n_1 - n_2)/n_1$$

This expression is easier to remember too.

It is important to remember that NA represents the acceptance angle. A large NA represents a large acceptance angle which also means a large Δ represents a large difference in the core and cladding refractive indices. On the face of it, it might appear that it is desirable to have a large NA, but this will lead to complications as will be seen shortly.

5.4 Group delay in step-index fibers

An optical waveguide is used as a transmission medium for communicating information. In digital communication systems, the information is contained in a series of optical pulses called 'bits' using Pulsed Code Modulation (PCM) techniques. So the information carrying capability of the optical communication system will depend upon the ability to discern the optical pulses arriving at the other end of the fiber. In this context, it is important to make sure that the optical pulses propagating through the fiber-optic medium do not broaden or 'disperse'. Any pulse broadening or dispersion will give rise to an increase in Bit Error Rate (BER) in the communication system.

The optical energy from the source gets coupled and propagates through the fiber in the form of discrete modes. The electromagnetic wave theory explaining the modal propagation in a fiber is given in Section 5.5. For the time being, in order to understand the concept, it is sufficient to assume that the input optical energy is distributed into certain number of modes which travel through different distances before arriving at the receiver end as shown in Fig. 5.7.

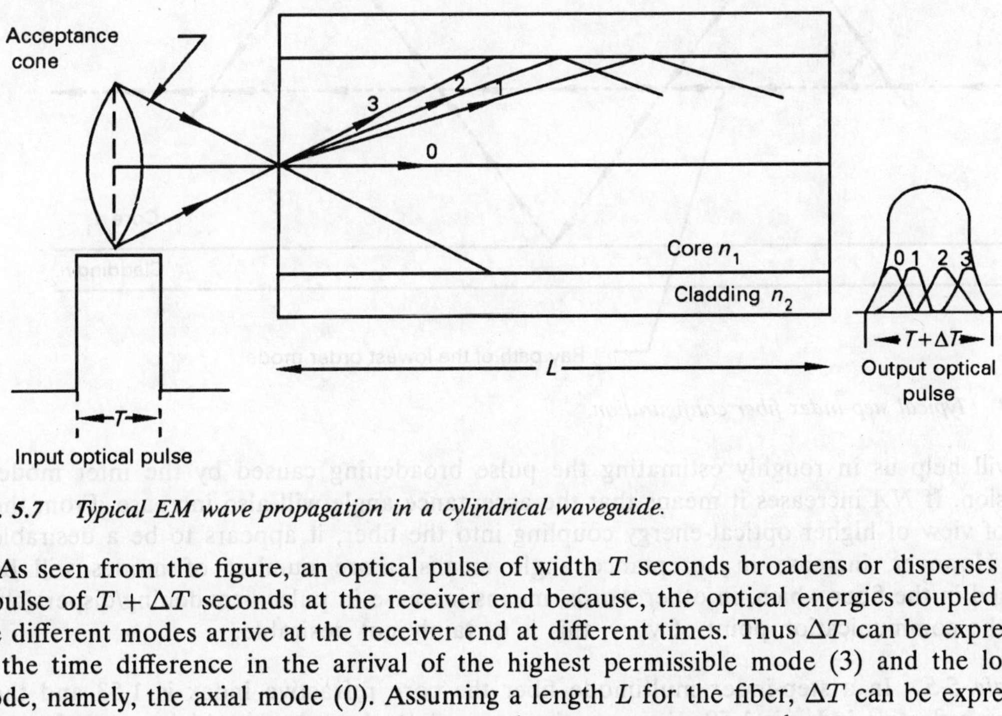

Fig. 5.7 *Typical EM wave propagation in a cylindrical waveguide.*

As seen from the figure, an optical pulse of width T seconds broadens or disperses into a pulse of $T + \Delta T$ seconds at the receiver end because, the optical energies coupled into the different modes arrive at the receiver end at different times. Thus ΔT can be expressed as the time difference in the arrival of the highest permissible mode (3) and the lowest mode, namely, the axial mode (0). Assuming a length L for the fiber, ΔT can be expressed as the difference in the travel time between these two extreme modes.

An estimate of the maximum difference in group delay ($\Delta\tau_g$) in an optical waveguide can be obtained by calculating the differential group delay between the highest and the lowest order modes.

$$\Delta\tau_g = \Delta\tau_{gL} - \Delta\tau_{gH} \tag{5.11}$$

where τ_{gL} and τ_{gH} are the group delays of the lowest and highest order propagating modes respectively, as shown in Fig. 5.8. As per the geometry of the figure, it can written as

$$\Delta\tau_g = \frac{L\sec\phi}{c/n_1} - \frac{L}{c/n_1} \tag{5.12}$$

where the highest order mode part length is $L\sec\phi$ compared to the path length L of the lowest order mode. This can be approximately written in the form

$$\Delta\tau_g \approx \frac{L}{c}(n_2 - n_1) \tag{5.13}$$

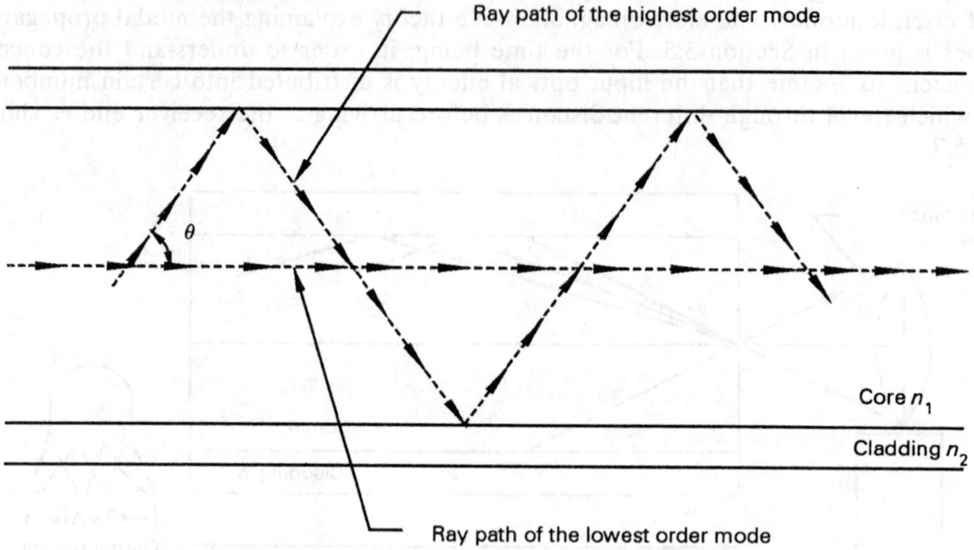

Fig. 5.8 Typical step-index fiber configuration.

This will help us in roughly estimating the pulse broadening caused by the inter model dispersion. If NA increases it means that the acceptance angle will also increase. From the point of view of higher optical-energy coupling into the fiber, it appears to be a desirable thing. However, increase in acceptance angle means, more number of modes will be sustained in the fiber which, in other words, means increase in pulse broadening/dispersion. From the communication point of view, this is certainly not desirable.

Example 5.5 In a step-index multimode fiber the core refractive index is 1.53 and the cladding reflective index is 1.50. Estimate the group delay between the highest and lowest order modes in a 3 km long fiber, if they propagate making an angle of 15° to each other.

Solution The group delay in the lowest order mode

$$\tau_{g2} = \frac{L_2}{c/n_{\text{core}}} = \frac{3000 \times 1.53}{3 \times 10^8} = 15.3\,\mu\text{s}$$

The group delay in the highest order mode

$$\tau_{g1} = \frac{L_1}{c/n_{\text{clad}}} = \frac{3000 \times 1.53 \times \sec(15^\circ)}{3 \times 10^8} = 15.836\,\mu\text{s}$$

Therefore, the differential group delay is given by

$$\Delta\tau_g = \tau_{g1} - \tau_{g2} = 0.536\,\mu\text{s}$$

5.5 Waveguide equations

So far, optical beam propagation through the fiber has been analysed using the simplistic ray optics approach. Now a more rigorous analysis of the problem will be taken up using the electromagnetic wave approach. The optical beam is nothing but the electromagnetic (EM) radiation at optical frequencies. Consider these EM waves propagating through a dielectric medium in the form of a cylindrical fiber as depicted in Fig. 5.9.

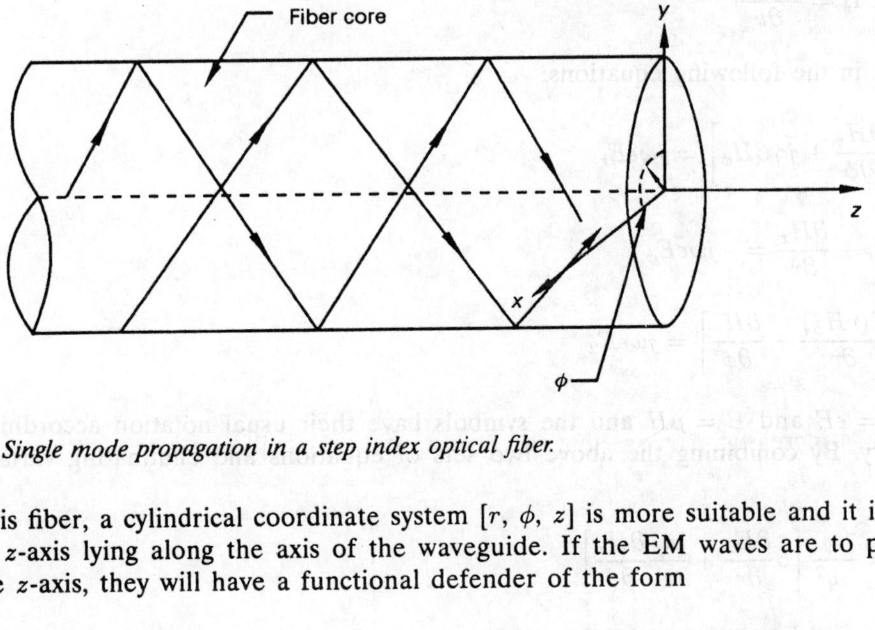

Fig. 5.9 Single mode propagation in a step index optical fiber.

For this fiber, a cylindrical coordinate system $[r, \phi, z]$ is more suitable and it is defined with the z-axis lying along the axis of the waveguide. If the EM waves are to propagate along the z-axis, they will have a functional defender of the form

$$E = E_0(r, \phi)\, e^{j(\omega t - z)} \tag{5.14}$$

$$H = H_0(r, \phi)\, e^{j(\omega t - z)} \tag{5.15}$$

These are harmonic in time t and propagate along the z-axis. The parameter β is the z-component of the propagation vector and will be determined by the boundary conditions on the electromagnetic fields at the core–cladding interface.

When Eqs.(5.14) and (5.15) are substituted into Maxwell's curl equations,

$$\text{curl } E = -\frac{\partial \bar{B}}{\partial t} \tag{5.16}$$

will yield

$$\frac{1}{r}\left[\frac{\partial E_2}{\partial \phi} + jr\beta E_\phi\right] = -j\omega\mu H_r \tag{5.17}$$

$$j\beta E_r + \frac{\partial E_2}{\partial r} = j\omega\mu H_\phi \tag{5.18}$$

$$\frac{1}{r}\left[\frac{\partial}{\partial r}(rE\phi) - \frac{-\partial E_r}{\partial \phi}\right] = -j\omega\mu H_z \tag{5.19}$$

in the cylindrical coordinate system, and for the magnetic field components,

$$\text{Curl } H = \frac{-\partial D}{\partial t} \tag{5.20}$$

will result in the following equations:

$$\frac{1}{r}\left[\frac{\partial H_z}{\partial \phi} + j\gamma\beta H_\phi\right] = j\omega\epsilon E_r \tag{5.21}$$

$$j\beta H_r + \frac{\partial H_z}{\partial r} = -j\omega\epsilon E_\phi \tag{5.22}$$

$$\frac{1}{r}\left[\frac{\partial (rH_\phi)}{\partial r} - \frac{\partial H_r}{\partial \phi}\right] = j\omega\epsilon E_2 \tag{5.23}$$

where $D = \epsilon E$ and $B = \mu H$ and the symbols have their usual notation according to the EM theory. By combining the above two sets of equations and eliminating variables, we obtain

$$E_r = -\frac{j}{q^2}\left[\beta\frac{\partial E_z}{\partial r} + \frac{\mu\omega}{r}\frac{\partial H_z}{\partial \phi}\right] \tag{5.24}$$

$$E_\phi = -\frac{j}{q^2}\left[\frac{\beta}{r}\frac{\partial E_z}{\partial \phi} + \mu\omega\frac{\partial H_z}{\partial r}\right] \tag{5.25}$$

$$H_r = -\frac{j}{q^2}\left[\beta\frac{\partial H_z}{\partial r} + \frac{\omega\epsilon}{r}\frac{\partial E_z}{\partial \phi}\right] \tag{5.26}$$

$$H_\phi = \frac{j}{q^2}\left[\frac{\beta}{r}\frac{\partial H_z}{\partial \phi} + \omega\epsilon\frac{\partial E_z}{\partial r}\right] \tag{5.27}$$

where $q^2 = \omega^2\epsilon\mu - \beta^2 = k^2 - \beta^2$. Substituting Eqs.(5.26) and (5.27) into Eq.(5.23), we get the EM wave equation in cylindrical coordinates

$$\frac{\partial^2 E_z}{\partial r^2} + \frac{1}{r}\frac{\partial E_z}{\partial r} + \frac{1}{r^2}\frac{\partial^2 E_z}{\partial \phi^2} + q^2 E_z = 0 \tag{5.28}$$

Substituting Eqs.(5.24) and (5.25) into Eq.(5.19), we get

$$\frac{\partial^2 H_z}{\partial r^2} + \frac{1}{r}\frac{\partial H_z}{\partial r} + \frac{1}{r^2}\frac{\partial^2 H_z}{\partial \phi^2} + q^2 H_z = 0 \tag{5.29}$$

It is interesting to note that both Eqs.(5.28) and (5.29) contain only either E_z or H_z. This appears to imply that the longitudinal components of E and H are uncoupled, and can be chosen arbitrarily provided they satisfy Eqs.(5.28) and (5.29). However, in general, coupling of E_z and H_z is required by the boundary conditions of the electromagnetic field components. If the boundary conditions do not lead to coupling between the field components, mode solutions can be obtained in which either $E_z = 0$ or $H_z = 0$. When $E_z = 0$, the modes are called transverse electric or TE modes, and when $H_z = 0$, they are called transverse magnetic or TM modes. Hybrid modes exist if both E_z and H_z are non-zero. These are designated as HE or EH modes, depending upon whether H_z or E_z, respectively, makes a larger contribution to the transverse field. The fact that hybrid modes are present in optical waveguides because of their dielectric nature makes the analysis more complex. But, in the case of microwaves, because hollow metallic waveguides are used, pure TE or TM modes can be found.

5.6 Wave equations for step-index fiber

We shall now extend the above analysis to find the guided modes in a step-index fiber. A standard and recommended mathematical procedure for solving Eqs.(5.28) and (5.29) is to use the separation of variables method, which assumes a solution of the form:

$$E_z = F_1(r)F_2(\phi)F_3(z)F_4(t) \tag{5.30}$$

where C is the proportionality constant.

As was already assumed, the time and z-dependent factors are given by

$$F_3(z)F_4(t) = e^{j(\omega t - \beta z)} \tag{5.31}$$

since the wave is sinusoidal in time and propagates in the z-direction. In addition, because of the circular symmetry of the waveguide, each field component must not change when the coordinate ϕ is increased by 2π. We will assume a periodic function of the form

$$F_2(\phi) = e^{j\tau\phi} \tag{5.32}$$

where τ is an integer that can be both positive and negative. Substituting Eqs.(5.30), (5.31) and (5.32) in Eq.(5.28), we get

$$\frac{\partial^2 F_1}{\partial r^2} + \frac{1}{r}\frac{\partial F_1}{\partial r} + \left(q^2 - \frac{\tau^2}{r^2}\right) F_1 = 0 \tag{5.33}$$

This is a well-known differential equation for Bessel function. Following the same steps, a similar equation can be derived for H_z too.

For a step-index fiber configuration, a homogeneous core of radius 'a' having a refractive index n_1 is considered. This is surrounded by a cladding of refractive index n_2 having a radius b where b tends to infinity as shown in Fig. 5.10.

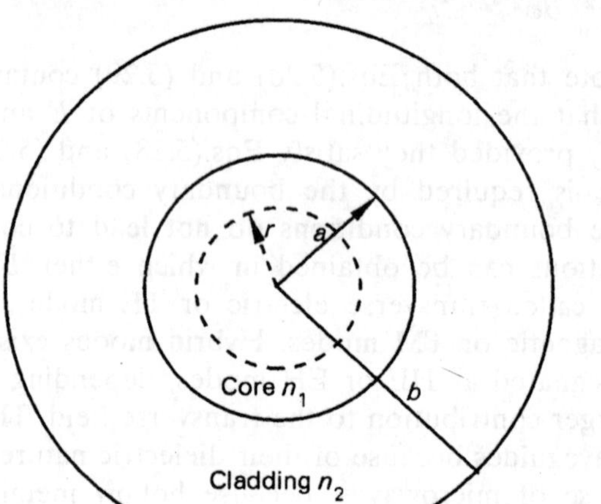

Fig. 5.10 Design curve for single mode fiber.

The reason for assuming an infinitely thick cladding is that the guided modes in the core set up an exponentially decaying field in the cladding must be insignificant, at the outer boundary of the cladding. Otherwise, the reflected fields from the outer boundary of the cladding might cause perturbation in the field pattern index of the core. In practice, however, b cannot be infinite and a realistic value is taken based on the theoretical and practical considerations.

Equation (5.33) must now be solved for the regions inside and outside the core. For the core region, the solutions for the guided modes must remain finite as $r \rightarrow 0$, whereas in the cladding, the solutions must decay to zero as $r \rightarrow \infty$. Thus, for $r < a$, the solutions are Bessel functions of the first kind in τ. For these functions, the common designation $J_\tau(\mu r)$ is used where $\mu^2 = k_1^2 - \beta^2$ with $k_1 = 2\pi n_1/\lambda$. The expression for E_z and H_z inside the core can then be written in the form:

$$E_z(r < a) = C_1 J_\tau(\mu r) e^{j\tau\phi} e^{j(\omega t - \beta z)} \tag{5.34}$$

$$H_z(r < a) = C_2 J_\tau(\mu r) e^{j\tau\phi} e^{j(\omega t - \beta z)} \tag{5.35}$$

where C_1 and C_2 are arbitrary proportionality constants.

Outside the core, the solutions of Eq.(5.33) are given by modified Bessel functions of the second kind in k. For these functions, the common designation is $k_\tau(\omega r)$, where $\omega = \beta^2 - k_2^2$ with $k_2 = 2\pi r_2/\lambda$. The expressions for E_z and H_z outside the core are, therefore,

$$E_z(r > a) = C_3 k_\tau(\omega r) e^{j\tau\phi} e^{j(\omega t - \beta z)} \tag{5.36}$$

$$H_z(r > a) = C_4 k_\tau(\omega r) e^{j\tau\phi} e^{j(\omega t - \beta z)} \tag{5.37}$$

where C_3 and C_4 are again arbitrary proportionality constants.

From the definition of the modified Bessel function of the second kind, it can be shown that $k_\tau(\omega r) \to e^{-\omega r}$ as $\omega r \to \infty$. Since $k_\tau(\omega r)$ must go to zero as $r \to \infty$, it follows that $\omega > 0$. This, in turn, implies that the propagation constant $\beta \geq k_2$ which represents a cut-off condition. The cut-off condition is the point at which a mode is no longer bound to the core regions. A second condition on β can be deduced from the behaviour of $J_\tau(\mu r)$. Inside the core, the parameter μ must be real for F_1 to be real and hence it follows that $k_1 \geq \beta$. The permissible range of β for the bound solution is, therefore,

$$n_2 k = k_2 < \beta < k_1 = n_1 k \tag{5.38}$$

where $k = 2\pi/\lambda$, and k is the free space propagation constant.

An important parameter connected with the cut-off condition is the normalized frequency V which is also called V number or V parameter. It is defined as

$$V^2 = (\mu^2 + \omega^2)a^2 = \left[\frac{2\pi a}{\lambda}\right]^2 (n_1^2 - n_2^2) \tag{5.39}$$

and is a one-dimensional number that determines the number of modes a fiber can support. When

$$V = \frac{2\pi a}{\lambda}\{n_1^2 - n_2^2\}^{1/2} = 2.405 \tag{5.40}$$

which is the value at which the lowest order Bessel function J_0 is zero, all modes are cut off excepting one single-mode. Single-mode fibers can, therefore, be designed and fabricated by making the core diameter sufficiently small and by having a small refractive index difference between the core and the cladding. From Eq.(5.40), it can be seen that single-mode propagation is possible for fairly large variations in physical core size a and core cladding refractive index differential. A typical single-mode fiber may have a core radius of 5 μm and a numerical aperture of 0.1 at a wavelength of 0.8 μm.

The V parameter can also be related to the number of modes M in a multimode fiber when M is large. An approximate relationship between these two can be deduced from the ray theory. A bunch of converging optical rays incident on the fiber end will be accepted by the fiber if it lies within the acceptance angle defined by the NA of the fiber:

$$NA = \sin\theta_{acc} = \{n_1^2 - n_2^2\}^{1/2} \tag{5.41}$$

For practical numerical apertures, $\sin\theta$ is small so that

$$\sin\theta_{acc} \approx \theta_{acc} \tag{5.42}$$

Then the solid acceptance angle Ω for the fiber is given by

$$\Omega = \pi\theta_{acc}^2 = \pi(n_1^2 - n_2^2) \tag{5.43}$$

For an electromagnetic radiation of wavelength λ emanating from a laser or a waveguide, the number of modes per unit solid angle is given by $2A/\lambda^2$, where A is the area the mode is leaving or entering. The factor 2 arises from the fact that the plane wave can have two polarization orientations. The area A in this case is the core cross-sectional area given by πa^2. Thus, the total number of modes M entering/leaving the fiber is

$$\begin{aligned} M &= \frac{2A}{\lambda^2}\Omega \\ &= \frac{2\pi^2 a^2}{\lambda^2}(n_1^2 - n_2^2) = \frac{V^2}{2} \end{aligned} \tag{5.45}$$

Example 5.6 Design a step-index fiber with a V-curve of 90 and a numerical aperture of 0.30. This fiber will have to be used in conjunction with GaAs laser diodes as the source with limit at 0.84 μm. Determine n_1, n_2, a and b of the fiber.

Solution As per the definition

$$V = 2\pi a/\lambda_0(n_1^2 - n_2^2)^{1/2}$$

$$NA = [n_1^2 - n_2^2]^{1/2}$$

For 0.84 μm, normally one would select fused silica as the core because it has minimum attenuation at this wavelength.

Thus $n_1 = 1.458$

$$\begin{aligned} n_2 &= [n_1^2 - NA^2]^{1/2} \\ &= [(1.458)^2 - (0.30)^2]^{1/2} = 1.426 \end{aligned}$$

$$\begin{aligned} a &= \frac{\lambda_0 V}{2\pi \times NA} \\ &= \frac{(0.84) \times 10^{-6} \times 90}{2 \times \frac{22}{7} \times 0.30} = 40.12\,\mu\text{m} \end{aligned}$$

To design the fiber fully, one would also need to specify the cladding radius b. The criterion for this is that the evanescent field in the cladding should be negligible at the cladding–air interface. This means

$$e^{\omega t} \to 0 \quad \text{as} \quad r \to b$$

As per the definition, $\omega^2 = R^2 - k_2^2$ with $k_2 = 2\pi n_2/\lambda$. This condition is approximately satisfied when $b \approx 63\,\mu\text{m}$.

5.7 Single-mode step-index optical fiber

The advantages of the single-mode fiber are obvious. The intermodal dispersion in a multimode fiber which places a restriction on the information capacity can now be eliminated. This is essentially achieved by restricting the core radius a so that there is only elbow room for one mode to propagate as shown in Fig. 5.11.

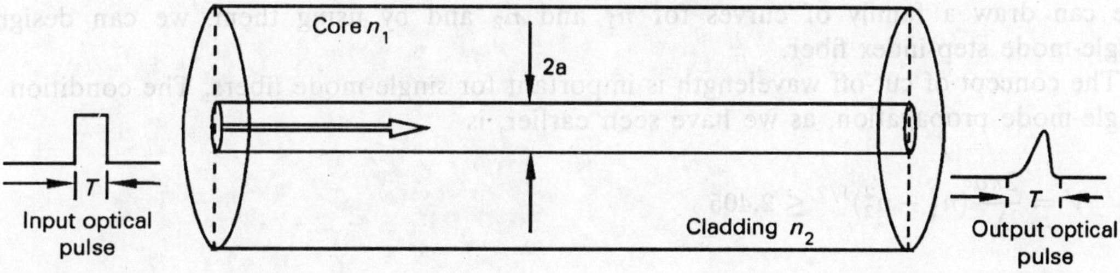

Fig. 5.11 Graded index fiber.

However, this has some disadvantages. Since the core diameter is drastically reduced, it would be somewhat difficult to couple the optical energy from the source into the fiber.

To develop the design equation for a single-mode fiber, let us rewrite Eq.(5.40) for the V number by putting

$$\Delta = \frac{n_1 - n_2}{n_1} \quad \text{or} \quad n_2 = n_1(1 - \Delta)$$

We get $V = \dfrac{2\pi a}{\lambda_0}(n_1^2 - n_2^2)^{1/2} = \dfrac{2\pi a}{\lambda_0} n_1 (2\Delta - \Delta^2)^{1/2} < 2.405$ (5.45)

For small Δ, $\Delta^2 \ll 2\Delta$ and can be neglected, and the above equation becomes

$$\frac{2\pi a}{\lambda_0}(2)^{1/2} n_1 (\Delta)^{1/2} = \frac{8.886 a n_1}{\lambda_0}(\Delta)^{1/2} < 2.405 \qquad (5.46)$$

The above condition can be achieved either by increasing the operating wavelength λ_0 by reducing the core radius a or by reducing the core refractive index n_1 or by reducing the refractive index differential between the core and the cladding Δ. Usually the wavelength λ_0

and the core refractive index are decided by extraneous considerations like the availability of suitable optical sources and the corresponding low-loss fiber optic dielectric method. However, the values for a and Δ can be engineered within certain limits. But the value of a cannot be drastically reduced since it would be increasingly difficult to couple light into the fiber. It is customary to use a small value for Δ, in the region of 0.002, but to reduce it any further would mean that the optical energy confinement in the core will not be effective.

Example 5.7 If we take a GaAs laser operating at 0.84 μm coupled to a single-mode step-index fiber of core refractive index of 1.458 and $\Delta = 0.002$, estimate the maximum permissible core radius for such a fiber.

Solution $V = 2.405 = \dfrac{8.886an_1}{\lambda_0}(\Delta)^{1/2}$

Therefore $a = \dfrac{2.405}{8.886 \times 1.458 \times 0.002} = 8.5\,\mu\text{m}$

We can draw a family of curves for n_1 and n_2 and by using them, we can design a single-mode step-index fiber.

The concept of cut-off wavelength is important for single-mode fibers. The condition for single-mode propagation, as we have seen earlier, is

$$V = \frac{2\pi a}{\lambda_0}(n_1^2 - n_2^2)^{1/2} \leq 2.405$$

This condition for single-mode operation will be satisfied for one particular wavelength and at any other higher wavelength. If V is greater than that particular value, the same fiber will support more than one mode. So the wavelength at which higher modes are cut off and the same optical waveguide behaves like a single-mode fiber, is referred to as the cut-off wavelength and is given by λ_c.

Example 5.8 For an optical waveguide having the following characteristics, calculate the cut-off wavelength λ_c: $a = 5\,\mu$m; $n_1 = 1.450$; $\Delta = 0.002$.

Solution If the V number of the fiber is greater than 2.405, the fiber will support more than one mode. We will calculate the cut-off wavelength λ_c for single-mode operation from the equation

$$V = 2.405 = \frac{2\pi a}{\lambda_c}[n_1^2 - n_2^2]^{1/2}$$

Solving for λ_c, we get

$$\lambda_c = \frac{2\pi \times 5\mu m}{2.405}\{(1.450)^2 - (1.447)^2\}^{1/2}$$
$$= 1.218\,\mu m$$

If λ_0 is less than 1.218 μm, the waveguide having a core radius of 5 μm will act as a multimode fiber.

5.8 Graded-index optical fibers

Though the single-mode step-index fibers can overcome the problem of pulse dispersion by restricting the number of propagating modes, it is only achieved at the expense of the optical power that is coupled into the fiber. An attempt is made in the case of graded-index fibers to overcome the pulse dispersion problem without decreasing the core radius. This is achieved by designing the core refractive index in such a way that it decreases from the centre of the core to the cladding which obeys the following function:

$$n(r) = n_1[1 - 2\Delta(r/\alpha)^2]^{1/2} \quad \text{for} \quad 0 \leq r \leq a \tag{5.47}$$

The cladding refractive index remains constant at n_2, as shown in Fig. 5.12.

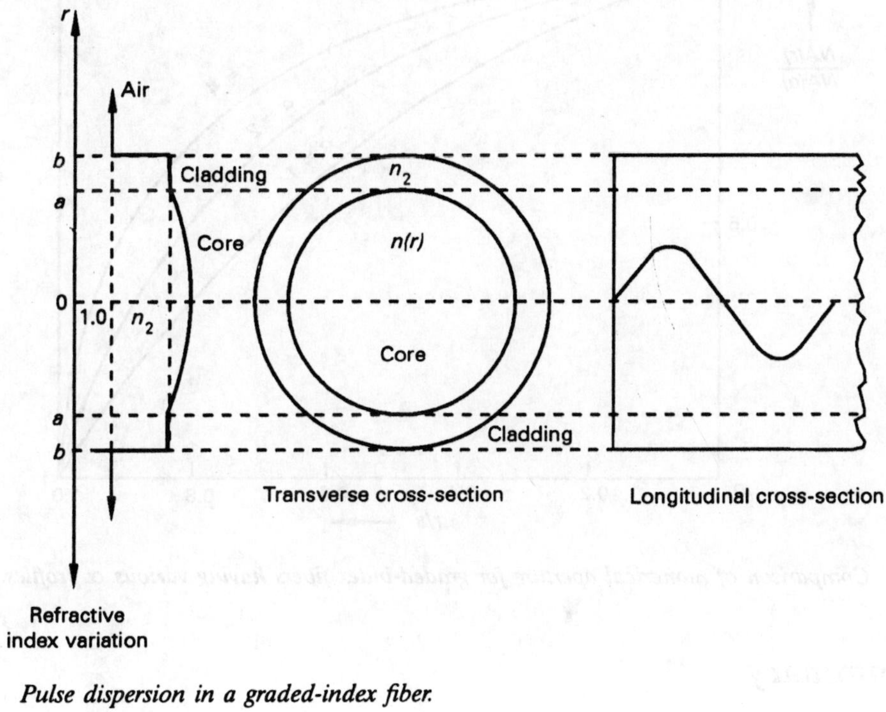

Fig. 5.12 Pulse dispersion in a graded-index fiber.

The dimensionless parameter α in Eq.(5.47) defines the shape of the index profile. The determination of NA for graded-index fibers is more complex than for step-index fibers. In graded-index fibers, NA is a function of position across the core end face. This is in contrast to the step-index fiber where NA is constant across the core. Geometrical optics considerations show that the light incident on the fiber core at position r will propagate as a guided mode only if it is within the local numerical aperture NA(r) at that point. The local numerical aperture is defined as:

$$\text{NA}(r) = [n^2(r) - n_2^2]^{1/2} = \text{NA}(0)[1 - (r/a)^\alpha]^{1/2}, \quad \text{for } r < a \tag{5.48}$$

where the axial numerical aperture NA (0), is defined as

$$\begin{aligned}\text{NA}(0) &= [n^2(0) - n_2^2]^{1/2} = (n_1^2 - n_2^2)^{1/2} \\ &= n_1(2\Delta)^{1/2}\end{aligned}$$

From Eq.(5.48), it is clear that the numerical aperture of a graded-index fiber decreases from NA(0) to zero as r moves from the fiber axis to the core–cladding boundary. A comparison of numerical aperture for fibers having various α profiles is given in Fig. 5.13.

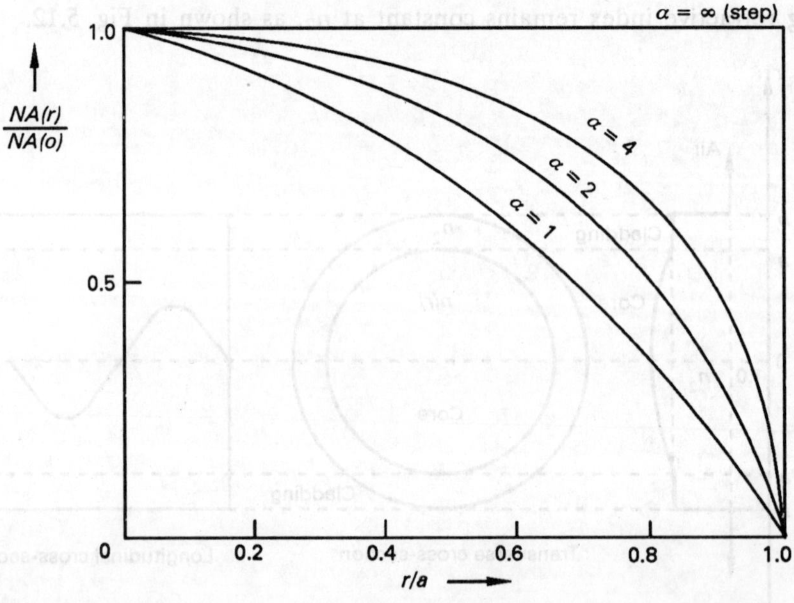

Fig. 5.13 Comparison of numerical aperture for graded-index fibers having various α profiles.

5.9 Summary

Optical fibers have come a long way after they were first conceived by Kao and Hockam in 1966. Narinder Singh Kapany developed and used fiber bundles with cladding for short-distance optical transmission. In 1970, Robert Maurer and his group at the Corning Glass Company in USA first produced communication grade low-loss fibers (loss number = 20 dB/km). Ever since, fiber losses have been steadily decreasing due to improved manufacturing technologies, and during the last few years fibers with losses as low as 0.2 dB/km are commercially available. This development means that transmission over a distance of more than 100 km (60 miles) is now possible, depending on the data rate without any repeaters. Now, high-intensity long-life semiconductor lasers may further increase this distance. Figure 5.14 gives a comparison of regions of propagation of multimode and single-mode step-index and graded index fibers. To begin with, fiber-optic communication systems have been installed since 1980 using multimode step-index fibers. In the mid 1980s graded-index optical fibers appeared on the scene because they offered higher bit-rates than multimode step-index optical fibers without compromising the coupling efficiency of the fiber to the source. However, because of the involved manufacturing process they were relatively expensive. In due course single-mode step-index fibers are favoured because of their high bit-rate capability. The smaller coupling efficiency of the fiber vis-a-vis multimode and graded-index fibers have been overcome by innovating coupling mechanisms from source to fiber.

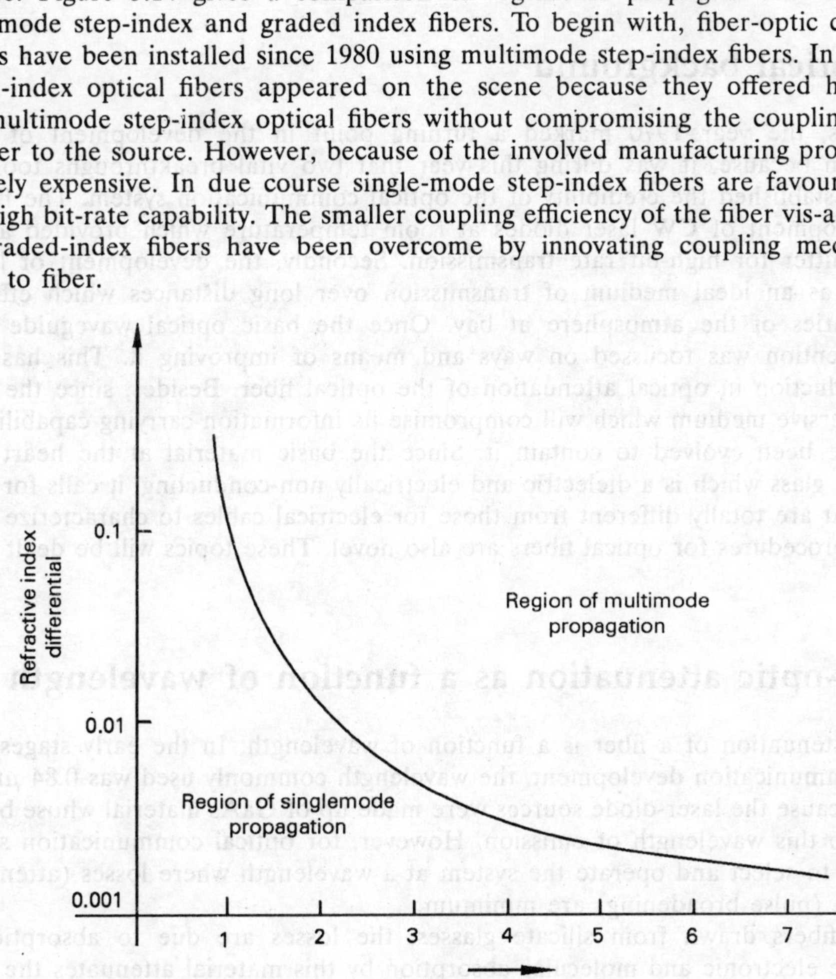

Fig. 5.14 Comparison of propagation regions in multimode and single-mode step-index fibers.

6 Optical fiber characterization and fabrication

6.1 Historical background

In many ways, the year 1970 marked a turning point in the development of optical communication because, it was during this year that two vital breakthroughs took place which firmly established the credibility of the optical communication system. The first one was the development of CW laser diodes at room temperature which provided an ideal optical transmitter for high-bit rate transmission. Secondly, the development of low-loss optical fibers as an ideal medium of transmission over long distances which effectively kept the vagaries of the atmosphere at bay. Once the basic optical waveguide was in place, the attention was focussed on ways and means of improving it. This has led to the drastic reduction in optical attenuation of the optical fiber. Besides, since the optical fiber is a dispersive medium which will compromise its information-carrying capability, new strategies have been evolved to contain it. Since the basic material at the heart of the optical fiber is glass which is a dielectric and electrically non-conducting, it calls for special techniques that are totally different from those for electrical cables to characterize it. The fabricational procedures for optical fibers are also novel. These topics will be dealt with in this chapter.

6.2 Fiber-optic attenuation as a function of wavelength

The optical attenuation of a fiber is a function of wavelength. In the early stages of the fiber-optic communication development, the wavelength commonly used was 0.84 μm. This was largely because the laser-diode sources were made up of GaAs material whose bandgap corresponds to this wavelength of emission. However, for optical communication systems, it is necessary to select and operate the system at a wavelength where losses (attenuation) and dispersion (pulse broadening) are minimum.

In optical fibers drawn from silicate glasses, the losses are due to absorption and scattering. The electronic and molecular absorption by this material attenuates the optical energy selectively by converting it to mechanical vibrations known technically as 'phonons'. This mechanical energy is then dissipated in the fiber in the form of heat. Besides losing useful optical energy, this can also change the refractive index profile of the fiber. The dominant absorptions are due to the infrared phonons absorption tail and the ultraviolet electron absorption tail by the constituent elements of the fiber as shown in Fig. 6.1. This results in a valley of low-loss region in the near infrared region of the operating wavelength. Besides, there are scattering losses due to Rayleigh and Mie scattering which are wavelength dependent. These scattering losses are due to directing of the optical energy away from the intended direction of propagation. Rayleigh scattering is a function of wavelength ($\propto \lambda^{-4}$), and reduces sharply in the infrared region. Mie scattering effects are, however, negligible because large-sized impurities in the fiber which cause this are usually avoided in the manufacturing process. Further, there will be sharp peaks in the absorption due to the presence of impurity ions like OH radical which diffuses into the fiber owing to its presence in ambient atmosphere. Thus, losses which result in the attenuation characteristics of the fiber as shown in Fig. 6.1 are

(a) Intrinsic material absorption loss
  (i) ultraviolet absorption tail
  (ii) infrared absorption tail
(b) Absorption loss due to impurity ions
(c) Rayleigh scattering loss
(d) Waveguide scattering loss
(e) Microbending loss.

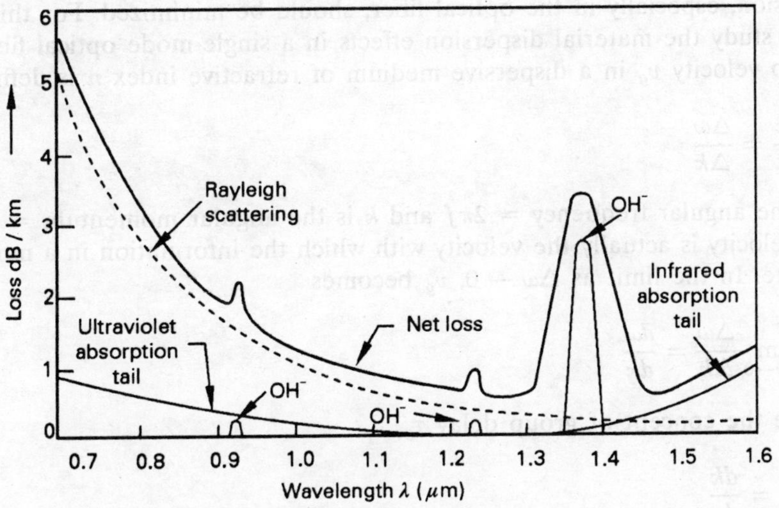

Fig. 6.1 The attenuation characteristics of an optical fiber (in dB/km) as a function of wavelength.

As a result, optical windows which are most suitable for operating the optical communication link will emerge. These are at 1.30 μm and 1.55 μm and the fiber attenuates much less optical energy at these wave lengths compared to any other wavelengths. Thus, it is advantageous to operate the link at these wavelengths. These wavelengths are referred to as second and third optical wave windows compared to the first optical window which was being operated in the early years based on GaAs laser diodes (0.84 μm). However, the exploitation of these windows would mean that one has to develop special devices and technologies which are capable of operating at these particular wavelengths.

6.3 Pulse delay and distortion in a single-mode fiber

As we have seen earlier, there is a considerable delay introduced in the pulse due to transit time in long-haul fiber-optic communication links. Besides, the input pulse is beaten out of shape by the time it arrives at the other end of the fiber. This distortion is due to pulse broadening or dispersion due to the following three important factors

(a) Inter-modal dispersion
(b) Material dispersion
(c) Waveguide dispersion

These three dispersion effects decrease in their order of magnitude. For example in a step-index multimode fiber where inter-modal dispersion is predominant, the contribution by material and waveguide dispersion is rather insignificant. However, in a step-index single-mode fiber, where the inter-modal dispersion is negligible, material dispersion comes into focus while waveguide dispersion still remains small.

When a very high density optical communication system is to be designed, it is necessary to design a system which can handle very high bit rates. This, in turn, means that the pulse dispersion, especially in the optical fiber, should be minimized. For this reason, it is necessary to study the material dispersion effects in a single-mode optical fiber.

The group velocity v_g in a dispersive medium of refractive index n is defined as

$$v_g = \frac{dz}{dt} = \frac{\Delta\omega}{\Delta k} \tag{6.1}$$

where ω is the angular frequency $= 2\pi f$ and k is the angular momentum $= 2\pi/\lambda = n\omega/c$. The group velocity is actually the velocity with which the information in a modulated wave will propagate. In the limit as $\Delta\omega \to 0$, v_g becomes

$$v_g = \lim_{\Delta\omega\to 0} \frac{\Delta\omega}{\Delta k} = \frac{d\omega}{dk} \tag{6.2}$$

Let us define the concept of group delay τ_g

$$\tau_g = \frac{1}{v_g} = \frac{dk}{d\omega} \tag{6.3}$$

Calculating the group delay τ_g from Eqs.(6.2) and (6.3), we get

$$\tau_g = \frac{dk}{d\omega} = \frac{1}{c}\frac{d}{d\omega}(n\omega) \tag{6.4}$$

Differentiating Eq.(6.4), we get

$$\tau_g = \frac{1}{c}\left(n + \omega\frac{dn}{d\omega}\right) \tag{6.5}$$

For the sake of notational simplicity, a refractive index for the dispersive medium N_g is introduced which is defined as

$$N_g = n + \omega \frac{dn}{d\omega} = \text{group refractive index} \tag{6.6}$$

Then $\tau_g = \dfrac{N_g}{c}$ (6.7)

and the group velocity then becomes

$$v_g = \frac{c}{N_g} \tag{6.8}$$

Since in the optical portion of the electromagnetic spectrum we often use wavelength rather than frequency, the equation for N_g can be rewritten as

$$N_g = n + \omega \frac{dn}{d\omega} = n + \omega \frac{dn}{d\lambda_0}\left(\frac{d\lambda_0}{d\omega}\right) \tag{6.9}$$

where $\lambda_0 = \dfrac{c}{f} = \dfrac{2\pi c}{\omega}$

$$\therefore \quad \frac{d\lambda_0}{d\omega} = -\frac{\lambda_0}{\omega}$$

Substituting the value of ω above in Eq.(6.9), we get

$$N_g = n - \lambda_0 \frac{dn}{d\lambda_0} \tag{6.10}$$

Note that the above equation reduces to $N_g = n$ in the case of non-dispersive media, where the refractive index is a constant, independent of the wavelength.

The optical transmitter, for example a laser diode (LD), emits radiation in a range of wavelength ($\Delta\lambda$) as shown in Fig. 6.2. Its relative refractive spectral bandwidth is defined as

relative spectral bandwidth of source $= (\Delta\lambda)/\lambda_0$

where $\Delta\lambda$ is the spread of wavelength around the centre frequency. Now, the corresponding spread in the arrival times of the optical energy can be written as

$$\Delta\tau_c = \frac{L}{c} N_g(\lambda_1) - \frac{L}{c} N_g(\lambda_2) \tag{6.11}$$

$$= \frac{L}{c} \frac{dN_g}{d\lambda} (\Delta\lambda) \tag{6.12}$$

Using the relation in Eq.(6.11) results in

$$\Delta\tau_c = \frac{L}{c} \lambda \frac{d^2 n_1}{d\lambda^2} \Delta\lambda \tag{6.13}$$

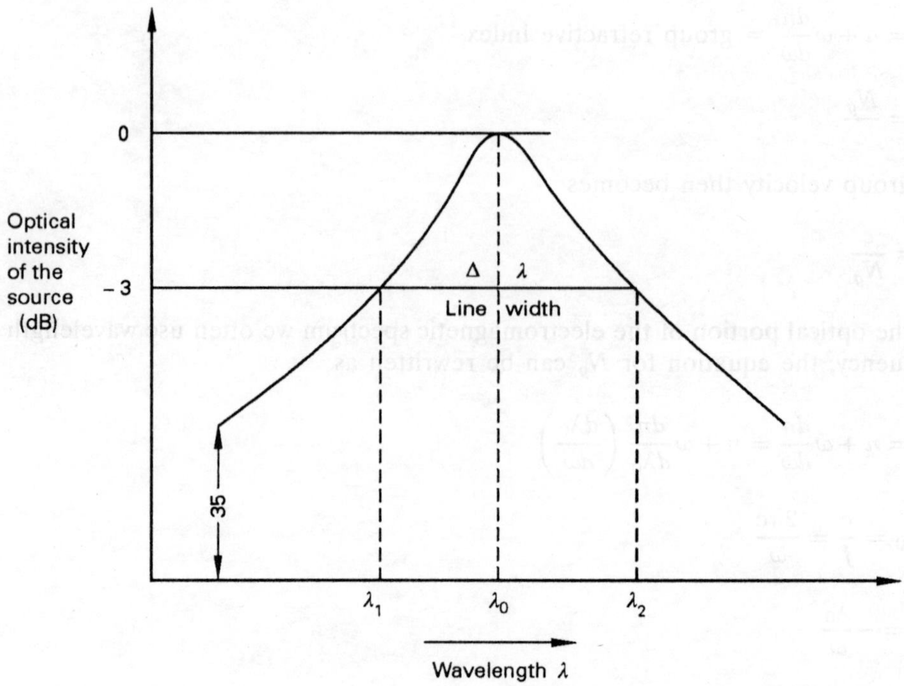

Fig. 6.2 The variation of optical intensity of the source as a function of wavelength of operation.

This can be rewritten in terms of the relative spectral bandwidth of the source as

$$\Delta\tau_c = \frac{L}{c}\frac{\Delta\lambda}{\lambda_0}\lambda_0^2\frac{d^2n_1}{d\lambda^2} \tag{6.14}$$

This is also known as chromatic dispersion because different wavelength components of the optical beam travel with different group velocities through the fiber. Assuming fused silica as the fiber material and with the help of its refractive index versus wavelength profile, we can plot $d^2n/d\lambda^2$ as a function of the wavelength as shown in Fig. 6.3. From the figure, it can be seen that the effect of material (chromatic dispersion) can be reduced to zero by operating at a wavelength in the vicinity of 1.3 μm. It may be recalled from the attenuation curve in Fig. 6.4, that there is a valley in the attenuation around this wavelength. Therefore, the second-generation fiber-optic communication links started adopting this wavelength by developing suitable devices for this purpose.

Example 6.1 Estimate the material/chromatic dispersion of a fused silica optical fiber of 1 km length for which $\lambda_0^2(d^2n/d\lambda^2)$ at 0.84 μm is 0.020. The GaAs source operates at 0.84 μm and has a linewidth of 1 nm.

Solution

Given, $\lambda = 0.84\,\mu\text{m} = 0.84 \times 10^{-6}$ m

$\Delta\lambda = 1$ nm $= 10^{-9}$ m

$$\lambda_0^2 \frac{d^2n}{d\lambda^2} = 0.02$$

$$L = 1 \text{ km} = 10^3 \text{ m}$$

$$c = 3 \times 10^8 \text{ m/s}$$

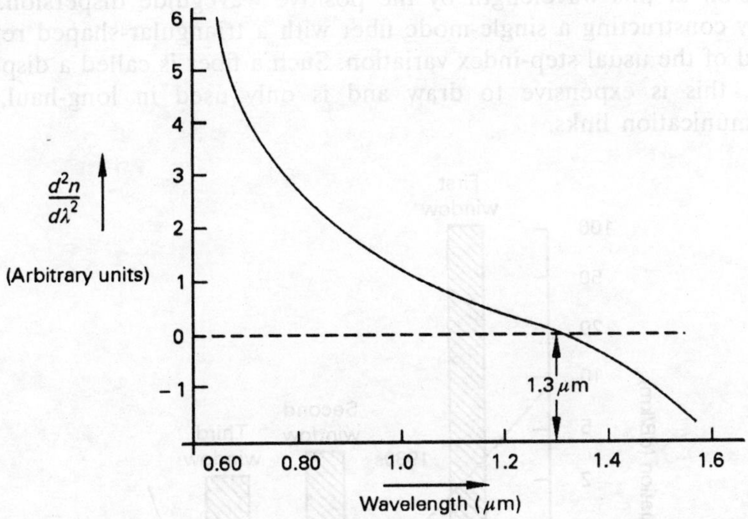

Fig. 6.3 *The variation of $d^2n/d\lambda^2$ as a function of the wavelength.*

Therefore, the chromatic dispersion is given by

$$\Delta\tau_c = \frac{L}{c}\frac{\Delta\lambda}{\lambda_0}\lambda_0^2\frac{d^2n}{d\lambda^2}$$

$$= \frac{10^3}{3 \times 10^8} \times \frac{10^{-9}}{0.84 \times 10^{-6}} \times 0.02 = 80 \text{ ps}$$

Let us now turn our attention to waveguide dispersion, which is of least importance amongst the three dispersion mechanisms. However, it comes into play once the inter-modal dispersion and the material dispersion are eliminated. Waveguide dispersion is a result of the fact that the propagating characteristics of the mode are a function of the ratio between the core radius and the wavelength. An expression for the waveguide dispersion $\Delta\tau_\omega$ is given by

$$\Delta\tau_\omega = \frac{L}{c}\frac{(\Delta\lambda)}{\lambda}(n_2 - n_1)D_\omega V \tag{6.15}$$

where D_ω is a dimensionless dispersion co-efficient which is a function of V, the wavenumber of the fiber. For a single-mode fiber using a GaAs laser source operating at 0.84μm with

$\Delta\lambda$ = 1 nm, τ_ω is estimated to be approximately 2 ps/mm of the fiber. This is much smaller than the material (chromatic) dispersion of the fiber and is usually ignored.

We have already seen from Fig. 6.4 that optical fibers have the lowest attenuation at 1550 nm. Thus, for long-haul communication links, this window has become important. However, at this wavelength, significant negative dispersion could limit the information carrying capability. This situation can be improved upon by compensating the negative material dispersion at this wavelength by the positive waveguide dispersion. This can be accomplished by constructing a single-mode fiber with a triangular-shaped refractive index variation instead of the usual step-index variation. Such a fiber is called a dispersion-shifted fiber. However, this is expensive to draw and is only used in long-haul, high-density fiber-optic communication links.

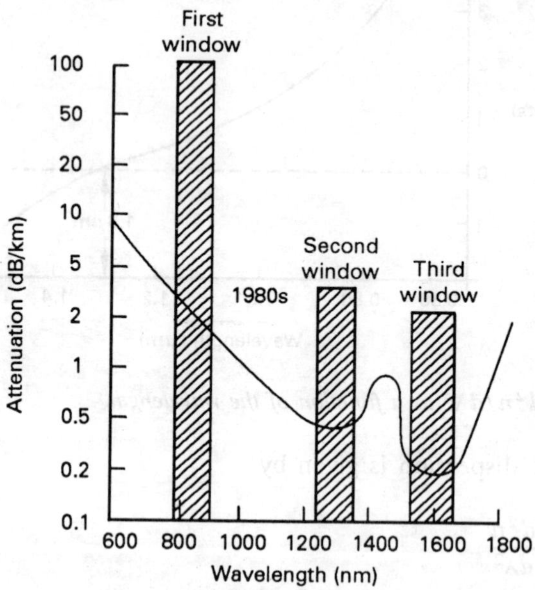

Fig. 6.4 The attenuation characteristics showing optical windows for long-haul optical communication links.

6.4 Fiber optic characteristics

The characteristics of glass, used for the fabrication of optical fibers are different from most solids found in nature. Many solids are crystalline in nature, made up of atoms positioned in precise predetermined slots called the lattice structure. Glass, on the other hand, is an amorphous material consisting of a loosely connected network of atoms whose characteristics can be modified. Thus, there is no precise melting point (temperature) for glass as in the case of crystalline materials. The viscosity of molten glass can vary smoothly over many orders of magnitude of temperature range (of many hundreds of degrees). This characteristic makes it an ideal candidate for fiberizing this material. To the pure fused silica glass (SiO_2), selected impurities called dopants are added. These impurities will alter the refractive index in a pre-determined way, as shown in Fig. 6.5, without unduly affecting the transparency characteristics of the material. The addition of the dopants to alter the refractive index profile of pure silica will also change the viscosity and the thermal expansion coefficients. The resulting difference in the coefficient of thermal expansion between the core and the cladding is important, since it causes the development of relative thermal expansion stresses within the fiber. These stresses can be harmful but they can be effectively utilized in a beneficial way to strengthen the fiber. It is desirable to have a cladding glass with smaller expansion coefficient as it will lead to axial compressive stresses in the cladding zone and assists in strengthening the fiber. Thus, it is important to understand that thermal expansion coefficients and viscosities of the core and the cladding glasses should be closely matched to ensure a good quality fiber.

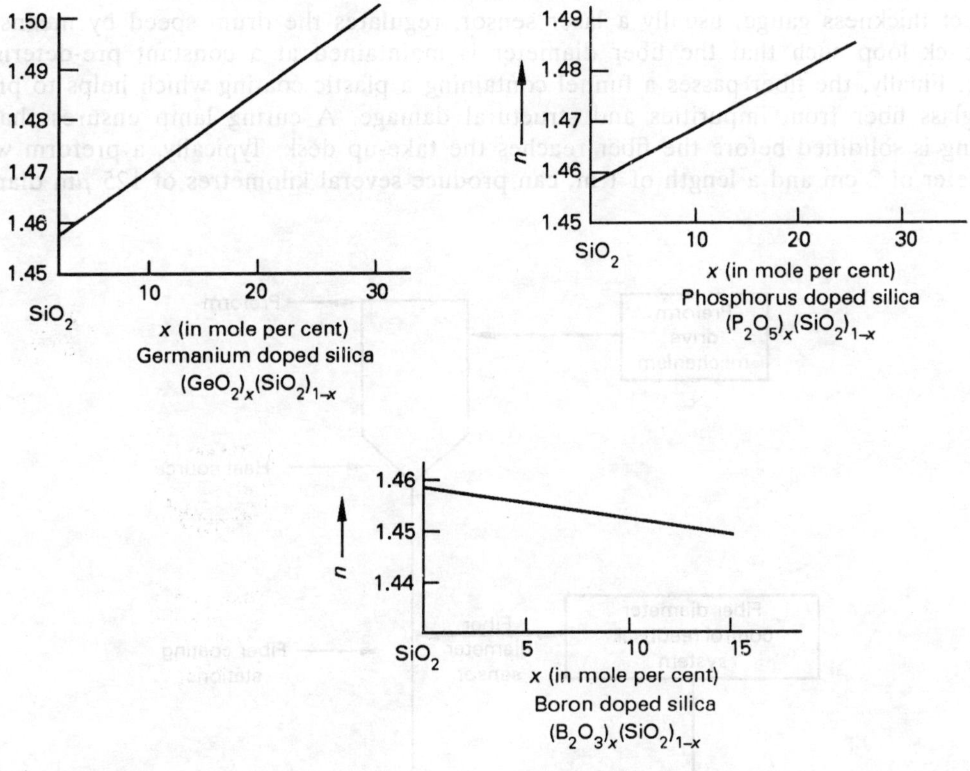

Fig. 6.5 The index of refraction for doped silica glass.

High-quality (pure) glass in the form of a rod, referred to as preform, consisting of a cladding surrounding a higher-index core material, can be made by different techniques. These are within the purview of glass technologists and for communication engineers, it is enough to understand the general manufacturing process. Many of the processes developed by glass manufacturers like corning glass are mostly found in patent literature. The commonly used process is the modified chemical vapour deposition technique developed by Bell Laboratories. Here, glass layers of doped silica are formed on the inside of the surface of a pure silica tube. The tube is then heated to a high temperature at which the surface tension and/or external pressure cause the tube to shrink radially until the central hole is eliminated. The resulting solid rod is the *preform* from which the optical fiber is drawn.

Once the preform is made, the next step is to design a suitable fiber drawing machine. This differs from one manufacturer to another. A basic sketch of a fiber drawing machine is given in Fig. 6.6. A clamp at the top of the fiber pulling tower holds the preform in place and drives the preform slowly but gradually downwards as the fiber is drawn. A heating furnace, located close to the bottom end of the preform, softens it to form a tip from which the fiber is drawn. Since the take-up located at the bottom of the tower, on which the fiber is wound, rotates, it determines the thickness of the fiber. So a non-contact thickness gauge, usually a laser sensor, regulates the drum speed by means of a feedback loop such that the fiber diameter is maintained at a constant pre-determined value. Finally, the fiber passes a funnel containing a plastic coating which helps to protect the glass fiber from impurities and structural damage. A curing lamp ensures that the coating is solidified before the fiber reaches the take-up desk. Typically, a preform with a diameter of 2 cm and a length of 1 m, can produce several kilometres of 125 μm diameter fiber.

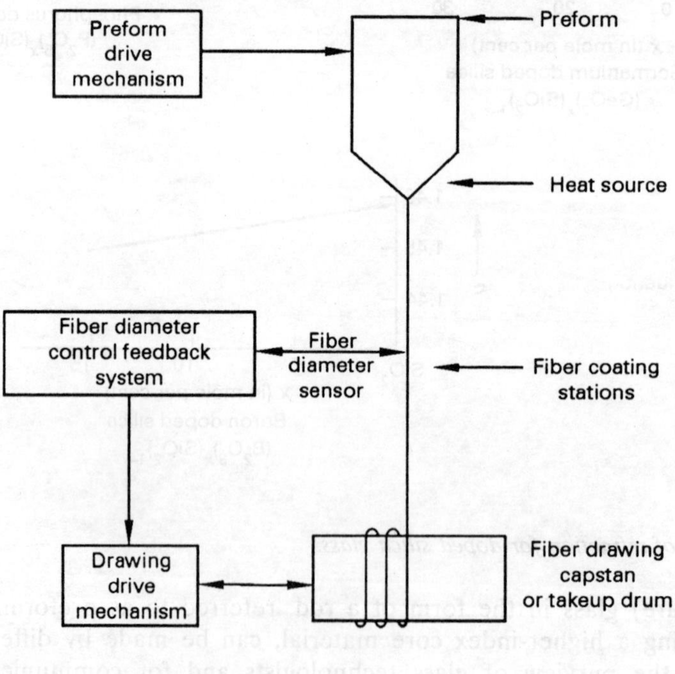

Fig. 6.6 The basic fiber drawing machine.

The optical fiber thus produced is too small and brittle for field use. It has to be strengthened by several layers of jackets before it can be deployed in the field. A typical single-fiber optical cable is shown in Fig. 6.7. The fiber will be further supported by Kevlar strength members and Teflon coatings just as in the case of copper cables. A fiber-optic cable ready for field application looks almost like any other metallic cable from outside. For example, the fiber-optic cable developed by the Bell Telecommunication Laboratories (BTL) and used in Bell's Chicago experiment during the 1970's is shown in Fig. 6.8. Another important aspect of fiber-optic communication is that its attenuation is virtually unaffected by the modulating frequency as compared to the conventional co-axial cables (Fig. 6.9).

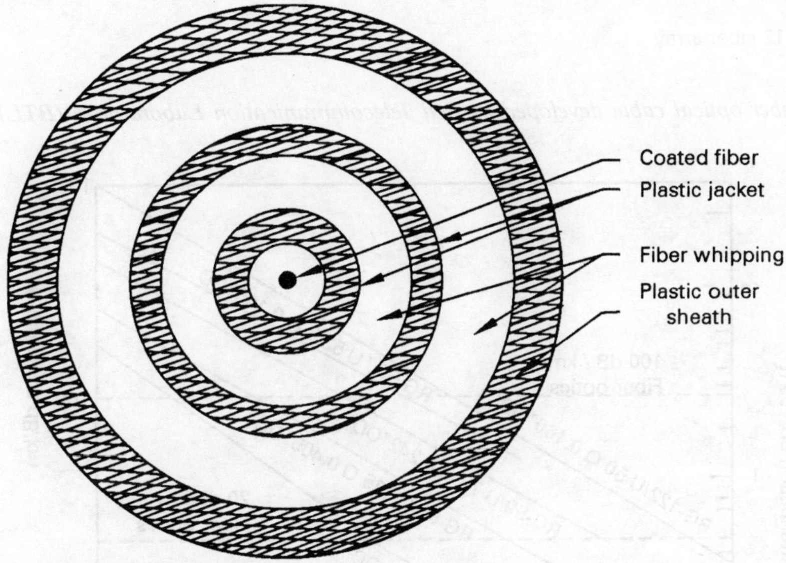

<u>General characteristics</u>

Attenuation	<15 dB/km
Bandwidth	Up to 400 MHz*
Overall diameter	5 mm
Weight	22 kg/km

<u>Mechanical capabilities</u>**

Bending radius	200 mm
Tensile load	200 N
Crushing load	4000 N/cm
Impact (0.14 kg-m, r=5 mm)	200 strokes
Flexing (±180°, r=50 mm	30,000 cycles

* "Step" or "Graded" fibers

** For rated optical characteristic and fiber strain <0.25%

Fig. 6.7 A typical single fiber optical cable.

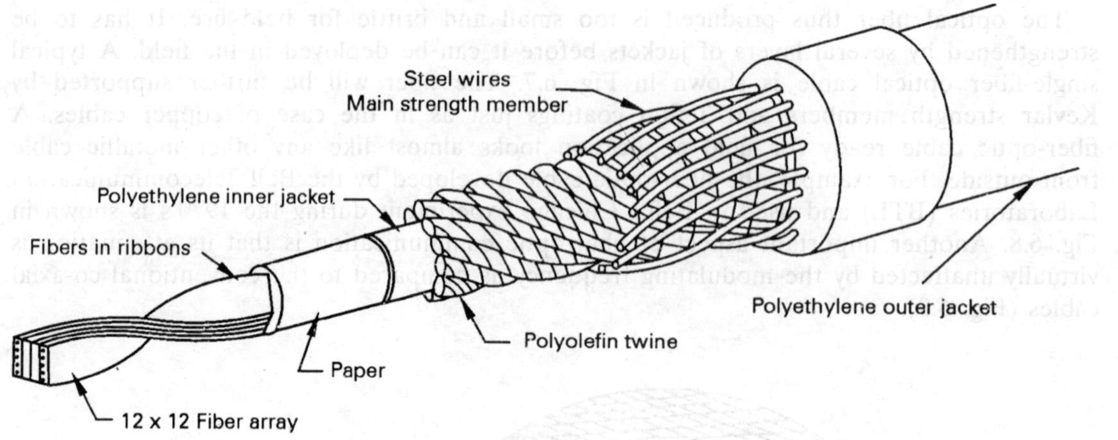

Fig. 6.8 The fiber optical cable developed by Bell Telecommunication Laboratories (BTL).

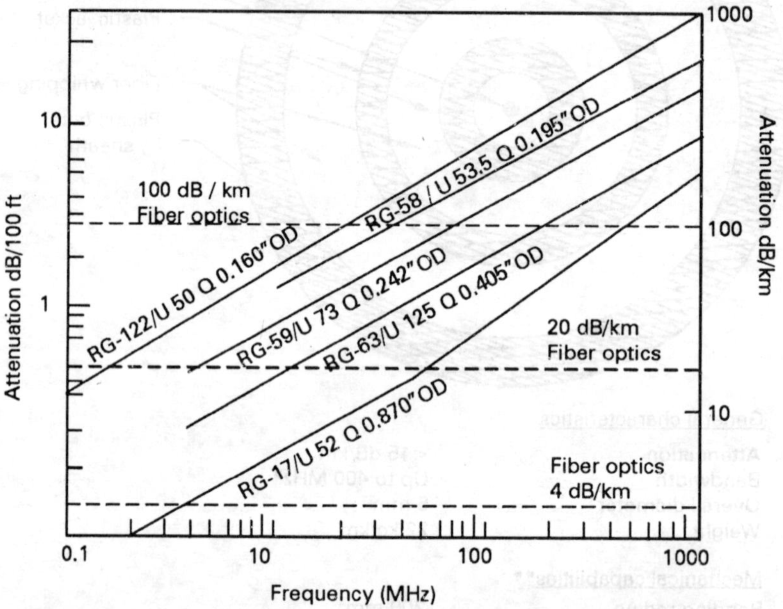

Fig. 6.9 Comparison between optical fiber and metal coaxial cable attenuations.

6.5 Connectors, couplers and splicing

Since the optical-fiber dimensions are relatively small, it requires a lot of ingenuity to couple-in and couple-out optical energy into and from the fiber. Further, since the fibers cannot be manufactured to unlimited lengths, the need to connect one fiber to another in the field either temporarily or permanently will arise. Though the LD and the LED have emerged as main optical sources for fiber-optic communication, the output profiles of optical energy from them are not uniform. As such, it was found necessary to introduce the fiber right into the source close enough to the active region so that there are better chances of coupling more energy into the fiber than otherwise. This resulted in a Burrus LED as shown in Fig. 6.10. This is also referred to as pig-tail configuration. At this point, it is important to note that as a rule, LDs which have higher radiance than LEDs will be able to couple-in more optical energy into the fiber.

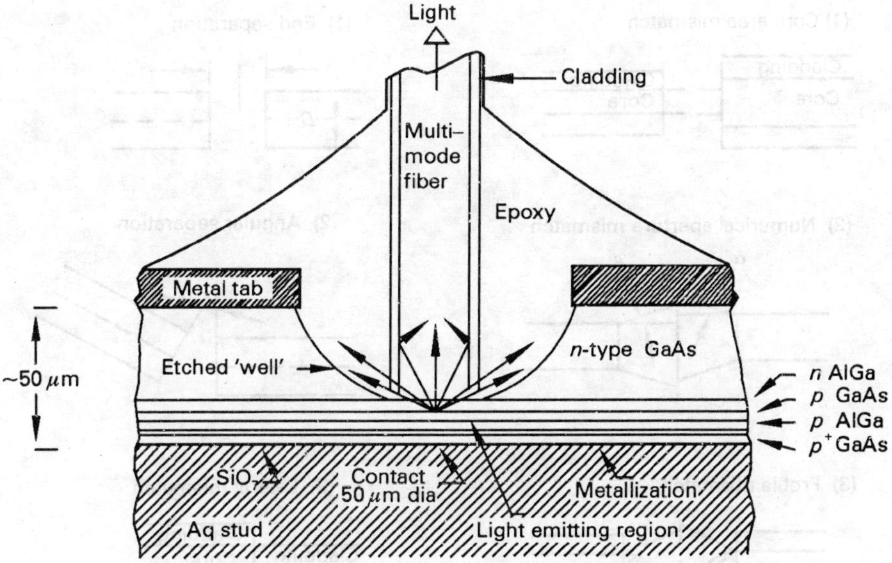

Radiance	100 W / sr / cm^2
Modulation B.W.	~200 MHz
Line width	350 Å
Power conv. eff.	1–3%
Lifetime	~100,000 hrs

$$P_0 = BA_s\,\Omega$$

Where $\Omega = 2\pi n_1^2\,\Delta, P_0 = 2\pi BA_s n_1^2\,\Delta$

Fig. 6.10 Burrus type LED.

When a fiber is cut, it acts like an open-ended waveguide. The near-field and far-field patterns from such a waveguide in the case of single-mode and multimode fibers have been well studied and understood. For fiber-optic connector design purposes, the radiation from such optical waveguides can be modelled with sufficient accuracy. Then the optical energy launched into the connecting fiber can be computed according to the principles of field matching. The geometry of the core and the light distribution within the core of the fiber govern the dimensional precision required for the fiber to be joined to another fiber to achieve a low-loss connection. The possible ways in which the losses can occur while joining one fiber to another are shown in Fig. 6.11. Many manufacturers have introduced several types of mountable/demountable fiber-optic connectors. They provide the required flexibility in routing/re-routing fiber-optic communication links. Notable among them is the multiple-ball concept for centering the fiber within the connector as shown in Fig. 6.12.

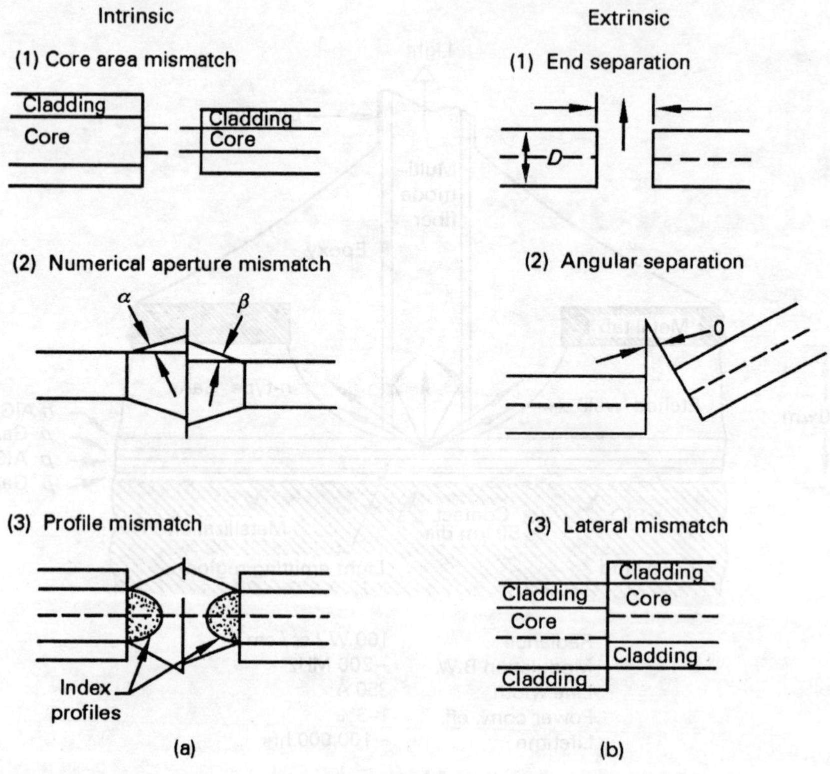

Fig. 6.11 Intrinsic and extrinsic coupling characteristics.

This design can hold the fiber in a particular position to a high degree of accuracy. It can be designed to accommodate fibers of varying cladding diameters of broad range and it is easy and quick to connect/terminate fibers which is an important consideration for field applications. When we wish to join two optical fibers permanently or semi-permanently we should use the splicing technique. There are two types of splicing techniques available: the fusion splicing (which is permanent) and the mechanical splicing (which is semi-permanent). These two splicing techniques are shown in Fig. 6.13.

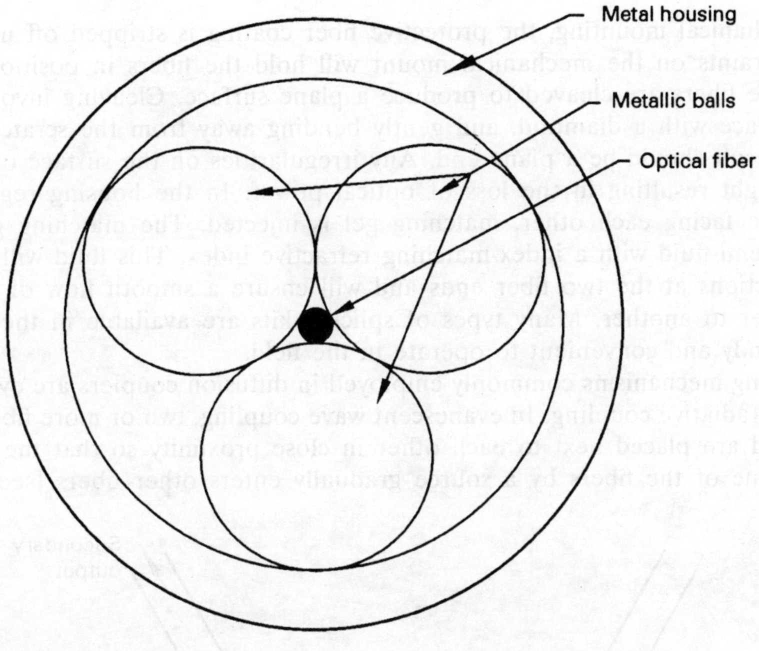

Fig. 6.12 *Multiple-ball arrangement for centering the fiber within a connector.*

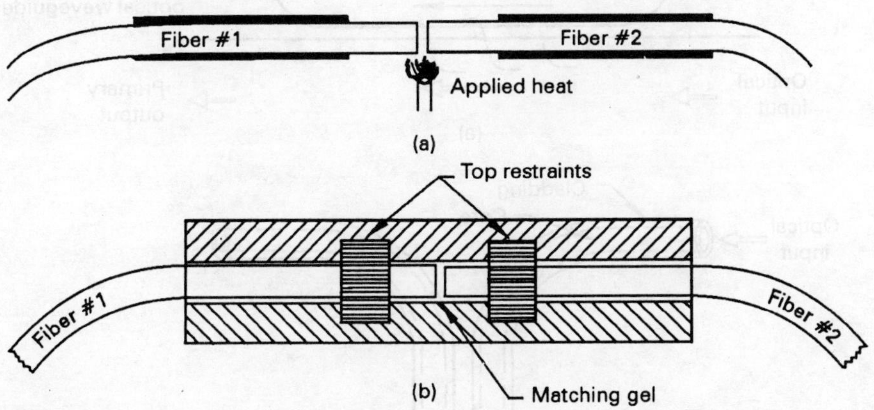

Fig. 6.13 *Splicing techniques (a) fusion splicing (b) mechanical splicing.*

In the case of fusion splicing, the two fiber ends are viewed through a microscope and are butted together using micropositioners. When they are correctly aligned, an electric arc is struck across the joint causing the two ends to melt and fuse. Inspection through the microscope reveals whether the joint is satisfactory or not; if not, it can be broken and remade. This technique results in a typical loss of 0.2 dB per splice, and so it is particularly attractive for use in long-haul routes. On the other hand, for short- and medium-haul routes, mechanical splicing is preferred, which provides far greater flexibility. Prior to inspection, into the mechanical mounting, the protective fiber coating is stripped off using a solvent. The top restraints on the mechanical mount will hold the fibers in position as shown in Fig. 6.13. The fibers are cleaved to produce a plane surface. Cleaving involves scratching the fiber surface with a diamond, and gently bending away from the scratch till the fiber snaps. The result should be a plane end. Any irregularities on the surface of the fiber end will scatter light resulting in the loss of optical power. In the housing region where the two fibers are facing each other, matching gel is injected. The matching gel is a highly transparent semi-fluid with a index-matching refractive index. This fluid will smoothen the Fresnel reflections at the two fiber ends and will ensure a smooth flow of optical energy from one fiber to another. Many types of splicing kits are available in the market today which are handy and convenient to operate in the field.

The coupling mechanisms commonly employed in diffusion couplers are evanescent wave coupling and radiative coupling. In evanescent wave coupling, two or more fibers which have to be coupled are placed next to each other in close proximity so that the optical power driven into one of the fibers by a source gradually enters other fibers [see Fig. 6.14(a)].

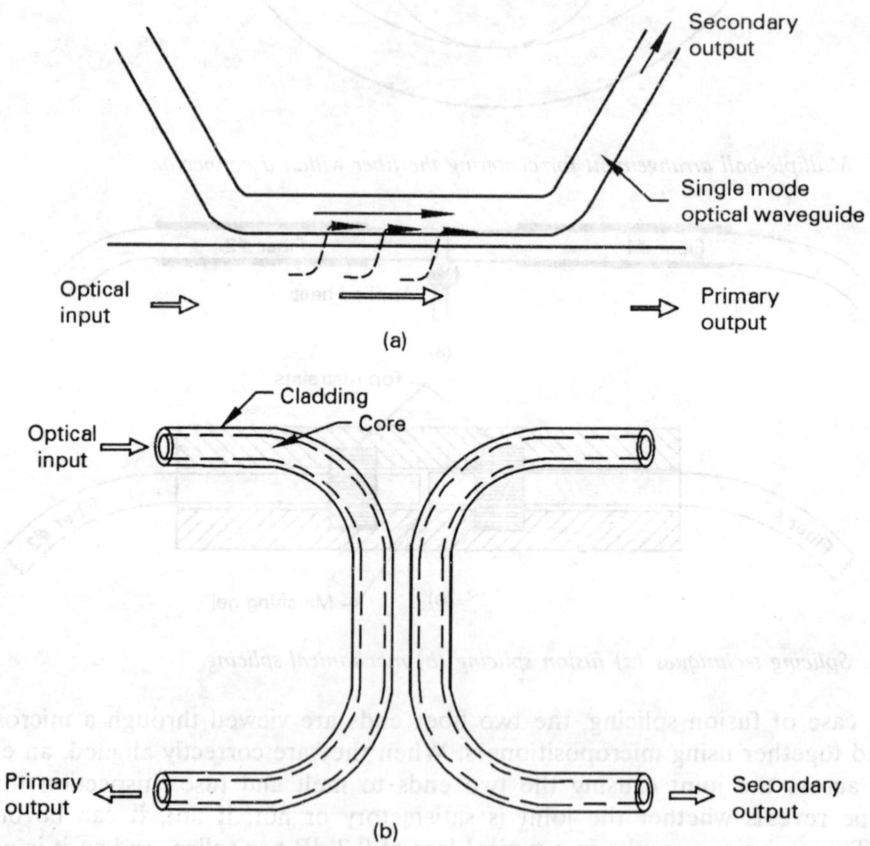

Fig. 6.14 *(a) Evanescent field coupler for single-mode fiber. (b) Radiation coupled bent fiber directional coupler.*

This phenomenon can be easily explained by the coupled equations of the electromagnetic theory. The transfer of optical energy from the primary fiber into the secondary fiber is through the evanescent field surrounding it. For effective transfer of energy, the connecting fibers are placed next to each other over a finite distance such that the evanescent field from the primary fiber builds up a propagation field in the secondary fiber. Such couplers are efficient for single-mode fibers whereas in multimode fibers, they are weak and mode-selective. This concept has been extended to designing and fabricating various types of diffusion couplers which find extensive-field applications in various situations.

In radiative coupling, the input fiber is bent at an acute angle in such a way that it radiates electromagnetic energy towards the receiving fiber. The receiving fiber is kept in the vicinity of this radiation field in such a way as shown in Fig. 6.14(b) that the energy is coupled into it optimally. For multimode fibers, this concept is extended by twisting the two fibers and fusing them together, as shown in Fig. 6.15. The light entering the coupler along one fiber core is radiated out of the core due to downtaper. This power is trapped in the cladding. Some of the trapped radiation then re-enters the core modes of each output fiber in the uptaper region. Coupling losses, defined as the power radiated out of the coupler, can be as low as 0.5 dB. Besides, there are many other types of optical couplers like area-splitting couplers, beam-splitter couplers, wavelength-selective couplers to mention only a few. The coupler (splicing) technology has advanced to such an extent in recent years that it can be effected in the field with ease even by a technician.

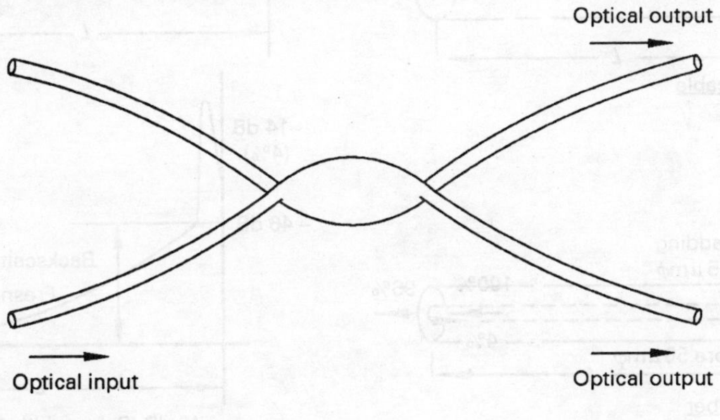

Fig. 6.15 Fused biconical taper coupler.

6.6 Fiber characterization

Since the medium used in optical fibers is dielectric and non-conducting, the conventional techniques used in the case of metallic cable characterization will not be useful for fiber-optic cables. The traditional way of measuring the attenuation of the optical fiber is to inject a known quantity of optical energy into one end of the fiber and then measure the optical output at the other end of it. From this, we can compute the fiber attenuation in terms of dB/km. However, this method will not help in locating any fault/discontinuity occurring at any place along the fiber. In order to overcome such problems, a new technique known as

Optical Time Domain Reflectometry (OTDR) has been conceived and developed. OTDR has proved to be a useful field-oriented loss-measurement method for determining the existence and location of fiber breaks and for measuring splice and transmission losses in an installed fiber system.

OTDR is essentially an optical pulse radar where an optical probe pulse (typically 5 to 10 ns half-power width) from the laser is injected into the test fiber which has to be characterized. As the probe pulse undergoes scattering continuously along the length of the fiber, it produces a low-level back scattered reflected signal at the same end of the fiber where the optical signal was injected. This can be compared with the electrical pulse which suffers a 100% reflection at the end of a metallic coaxial cable as shown in Fig. 6.16. The back scattered signals received at the input end of the fiber are detected by a photodetector and are displayed on a sampling oscilloscope which is referred to as the OTDR signature of the fiber. A block diagram of the OTDR experimental set-up is shown in Fig. 6.17. Typical numbers of the back scattered received signals (R_b) are given in Fig. 6.18 for graded-index (GI) and single-mode fibers. This will give us a feel of the processes and numbers connected with the performance of an OTDR. The major functions of OTDR are: (a) fault location, (b) optical loss per unit length, and (c) evaluation of splicing.

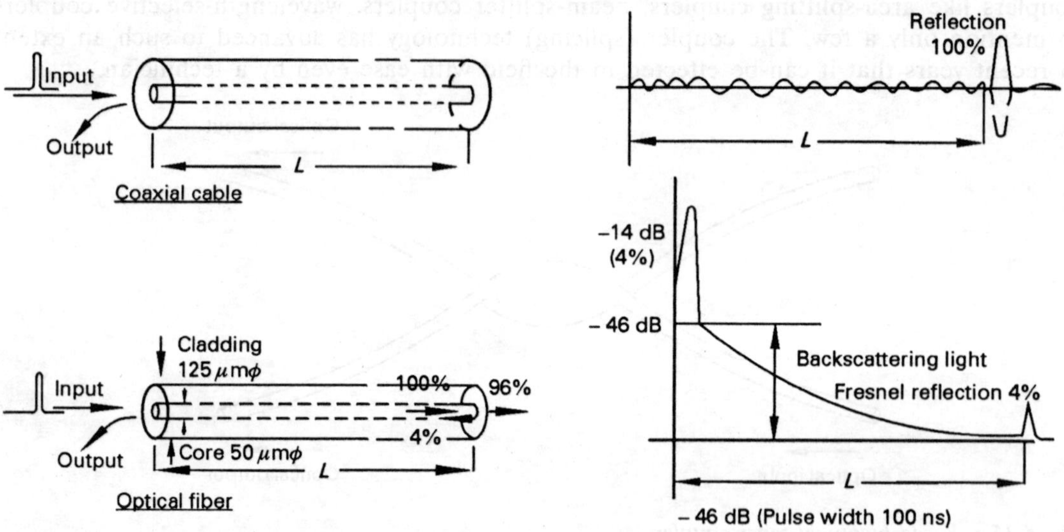

Fig. 6.16 Comparison of coaxial cable and optical fiber.

There are several applications of an OTDR, the most important of these being

(a) optical fiber cable manufacture,
(b) optical fiber cable installation,
(c) manufacture and repair of optical fiber transmission lines, and
(d) development and evaluation of optical fibers.

Though OTDR is the most important instrument for the characterization of an optical fiber, there are several other commercially available types of instruments developed for monitoring the performance of a fiber-optic communication system.

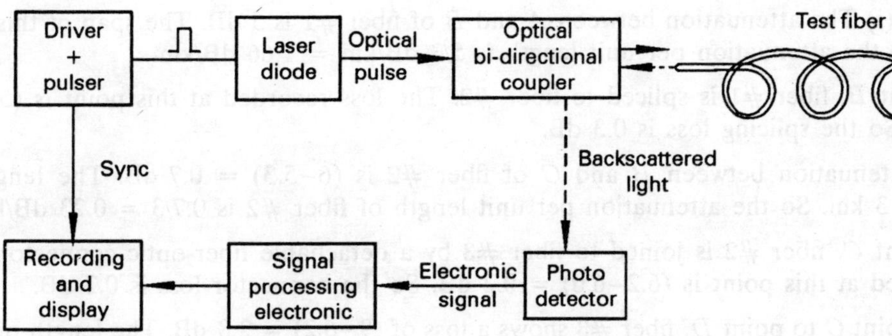

Fig. 6.17 Block diagram of OTDR.

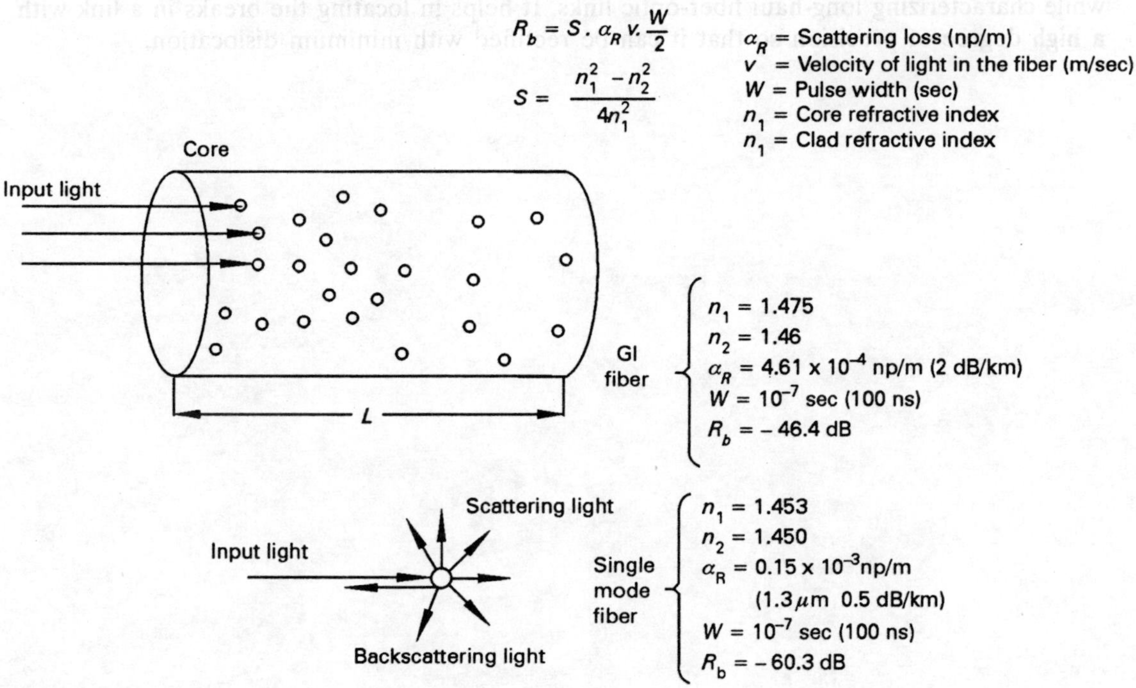

Fig. 6.18 The backscattering of light in an optical fiber.

Example 6.2 In the OTDR signature of a fiber-optic emitter given below, fiber #1 is spliced to fiber #2 whereas fiber #2 is connected to fiber #3 using a detachable fiber-optic connector. From the data given on the fiber, compute the attenuation per unit length of fibers #1, #2 and #3 and determine the splicing and connector losses. All the three fibers have equal lengths of 3 km each.

Solution (i) The attenuation between A and B of fiber #1 is 5 dB. The span of this fiber is 3 km. So the attenuation per unit length is 5/3 dB/km = 1.66 dB/km.

(ii) At point B, fiber #1 is spliced to fiber #2. The loss recorded at this point is (5.3–5) = 0.3 dB. So the splicing loss is 0.3 dB.

(iii) The attenuation between B and C of fiber #2 is (6–5.3) = 0.7 dB. The length of fiber #2 is 3 km. So the attenuation per unit length of fiber #2 is 0.7/3 = 0.23 dB/km.

(iv) At point C, fiber #2 is joined to fiber #3 by a detachable fiber-optic connector. The loss recorded at this point is (6.2–6.0) = 0.2 dB. So the connector loss is 0.2 dB.

(v) From point C to point D, fiber #3 shows a loss of (9–6.2) = 2.8 dB. The length of fiber #3 is 3 km. So the attenuation per unit length of fiber #3 is 2.8/3 dB/km = 0.93 dB/km.

In conclusion, the OTDR's main advantage lies in the fact that it is non-destructive in nature and requires access to *only one end* of the fiber. This is very important in the field while characterizing long-haul fiber-optic links. It helps in locating the breaks in a link with a high degree of precision so that it can be rectified with minimum dislocation.

7 Photodetectors and optical receivers

7.1 Historical background

The eye is perhaps the most sensitive natural photodetector. However, its spectral response depends on the wavelength—ranging between 0.4 and 0.7 μm—and its response is too slow. Photodetectors have found widespread application in common electronics in the form of light meters for cameras and cinema projectors much before the advent of optical communication. So, the basic groundwork on the development of photodetectors was already in place by the 1960s. It only needed to expand their capabilities for optical communication applications. The earliest photodetector was, perhaps, the selenium photoresistor which was successfully used by Alexander Graham Bell in his first optical communication experiment. However, it was found to be too slow for modern optical communication systems. Another photodetector which has found widespread application is the photocell which makes use of the photoemissive effect. The early photocells were based on the vacuum tube technology and are, therefore, not suitable for field use in optical communication systems. Only after the advent of semiconductor technology have compact, fast and efficient photodetectors been developed, which are suitable for application in optical communication.

7.2 Principles of photodetection

Photodetectors are at the heart of an optical communication receiver which is preceded by optics and succeeded by electronics. Thus, the efficiency of the photodetector will determine the overall performance of an optical communication system. In principle, a photodetector converts the optical energy falling on it into electrical signals which can be subsequently amplified and processed for extracting information.

Photodetectors fall into two classes: photon and thermal detectors. The latter is of little consequence for optical communication application because it is inherently very slow. Photon detectors can be further subdivided based on the physical principle of their operation: (a) photoemissive effect, (b) photoconductive effect, (c) photovoltaic effect and (d) photoelectromagnetic effect. The photoemissive effect in valves is the emission of electrons from a vacuum ray cathode when optical energy is made to fall on it. The other types of photodetectors can also be achieved using semiconductor materials.

There are several specifications which indicate the performance criteria of a photodetector. The following is the list of specifications adopted by most manufacturers of photodetectors:

(1) Quantum efficiency (η) It is defined as the ratio of the number of electrons emitted or generated by the photodetector to the average number of photons impinging on it. This parameter has no units since it is a ratio of numbers.

(2) Responsivity or sensitivity (R_0) It is defined as the ratio of the average photodetector current produced to the average incident optical radiation causing it. The photometric units of R_0 are amp/lumen, but at a specific wavelength it can be expressed as amp/watt.

(3) Spectral response It describes the way in which the quantum efficiency or responsivity of the photodetector changes as a function of the wavelength of optical radiation.

(4) Frequency response When the incident optical radiation is modulated sinusoidally, the electrical output from the detector decreases as the frequency of modulation is increased. The frequency response of the photodetector is often taken as the modulation frequency at which the output current of the photodetector decreases to half the peak value. The unit for this parameter is hertz (Hz).

(5) Impulse response When a perfect optical pulse with zero 'rise-time' and zero 'fall-time' infringes on a photodetector, the resultant electrical output pulse will have finite 'rise-time' and 'fall-time'. This is due to the capacitive aspect of the photodetector.

(6) Dark current (I_D) When the photodetector is totally shielded from any optical radiation, one would expect zero electrical output. However, in practice, one would still notice a small electrical current I_D in the photodetector output even under total 'darkness' conditions. This is referred to as the dark current. This parameter is measured in amperes.

(7) Noise equivalent power (NEP) Across the output terminals of the photodetector, random fluctuating voltage will appear even without an input–optical signal. The rms value of sinusoidally modulated optical radiation at 400 Hz with a reference bandwidth of 1 Hz which produces this equivalent output electrical signal is referred to as the noise equivalent power.

(8) Detectivity (D) It is the reciprocal of NEP.

$$D \equiv 1/\text{NEP}.$$

(9) D-star, $D^*(\lambda, f)$ It is defined as the detectivity multiplied by the square root of the difference of the detector area, A_D and the square root of the detector bandwidth, Δf.

$$D^*(\lambda, f) \equiv D[A_D \Delta f]^{1/2}$$

Specification of D^* requires knowledge of the modulation frequency and the detector temperature.

For the selection of a photodetector for optical communication for the best possible signal-to-noise ratio (SNR), the following considerations are very important.

1. The highest quantum efficiency for background radiation-limited operation.
2. The highest D^* (lowest NEP) for internal detector noise-limited operation.

Of course, added to this, the frequency/impulse response of the detector will have to be considered when one is planning a high bit-rate optical communication system.

7.3 Photoemissive detectors

The photoemissive detectors which convert the vacuum photodiode and photomultiplier tubes are given in Fig. 7.1. These are the forerunners to the present-day semiconductor photodetectors used in optical communication. The basic principle used in both these detectors is that the cathode emits electrons when it is exposed to optical radiation. The electrons emitted from the cathode are attracted and gathered by the anode to which a positive potential is applied. This results in a signal current through the load resistance R_L across which the signal voltage, v_s, is tapped. The maximum energy (kinetic) of an emitted electron is given by

$$E_e = hf - w \tag{7.1}$$

where f is the frequency of the optical radiation falling on the cathode and w is the work function of the cathode material which is the binding energy of the outermost electron. Only photons above a threshold frequency have energy larger than the work function of the material and, therefore, can cause electron emission. Alkali metals and certain semiconductor compounds have work functions that permit photoemissive operation for wavelengths of the order of 1 μm or less. Since the device has two electrodes in it, it is usually referred to as a photodiode.

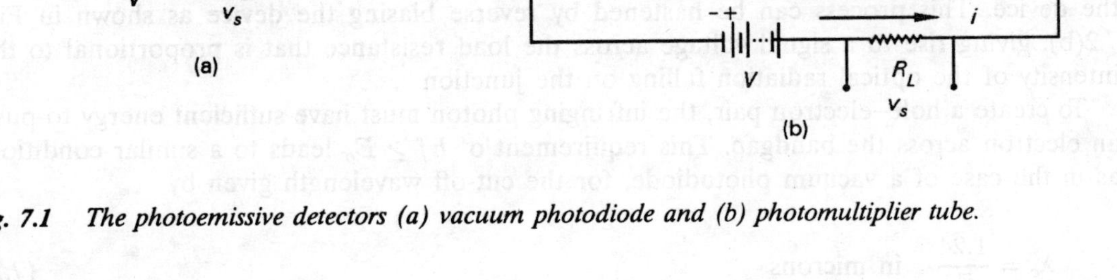

Fig. 7.1 The photoemissive detectors (a) vacuum photodiode and (b) photomultiplier tube.

Example 7.1 Cesium with a work function of 1.9 eV is a popular photoemissive material for making the cathode in a photodiode. Calculate the cut-off wavelength of the optical radiation it can detect.

Solution At cut-off wavelength λ_c

$$E_e = 0 = \frac{hc}{\lambda_c} - w$$

$$\text{Therefore, } \lambda_c = \frac{h \times c}{w} = \frac{6.626 \times 10^{-34} \times 2.998 \times 10^{8}}{1.9}$$

$$= 0.65\ \mu\text{m}.$$

This means that a photodiode with cesium cathode can only detect optical radiation of 0.65 μm or of shorter wavelength.

In the case of a photomultiplier tube (PMT), the internal gain is achieved by placing 'dynodes' in the electron path which result in secondary emission. The successive dynodes are operated at an increasing positive potential thus accelerating the electrons to cause secondary emission. A gain of 3 to 6 per stage is attainable by photomultiplication, yielding a total current gain for a typical nine-stage photomultiplier tube to be around 10^5 to 10^6.

The photoemissive detectors, however, suffer from many disadvantages which make them unsuitable for applications in optical communication. They are based on vacuum tube technology which makes them fragile and unfit for field applications. They require high operating voltages which are difficult to provide in the field. There is a finite transit time involved in the process of detection which makes it difficult for pulse operation. Besides, there is a capacitance effect due to the electrodes, which is of the order of a few tens of picofarads, that will restrict the bandwidth of an optical communication link.

7.4 Semiconductor photodetectors

Semiconductor devices, on the other hand, provide all the classical advantages of being small, rugged, fast, lightweight and capable of operating at low voltages. Thus, they are ideal candidates for field use in optical communication applications. The simplest form of semiconductor photodetector is an unbiased *pn* junction. When a photon impinges on the depletion region of a *pn* junction, it gives rise to a hole–electron pair as shown in Fig. 7.2(a). The hole and the electron tend to move in opposite directions under the influence of barrier potential set up across the junction due to diffusion, which drives current through the device. This process can be hastened by reverse biasing the device as shown in Fig. 7.2(b), giving rise to a signal voltage across the load resistance that is proportional to the intensity of the optical radiation falling on the junction.

To create a hole–electron pair, the infringing photon must have sufficient energy to push an electron across the bandgap. This requirement of $hf \geq E_g$ leads to a similar condition, as in the case of a vacuum photodiode, for the cut-off wavelength given by

$$\lambda_c = \frac{1.24}{E_g} \quad \text{in microns} \tag{7.2}$$

where E_g is expressed in electron volts.

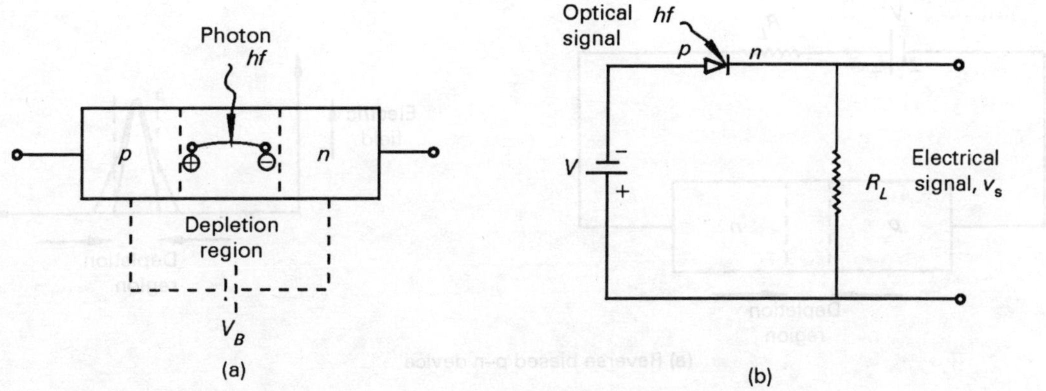

Fig. 7.2 (a) *p–n junction photodetector (b) semiconductor photodetector in operation.*

Example 7.2 Compute the cut-off wavelength for silicon (Si)-based and germanium (Ge)-based photodiodes. The energy gaps for Si and Ge are 1.1 eV and 0.67 eV, respectively.

Solution For Si, $\lambda_c = \dfrac{1.24}{1.1} = 1.1\ \mu m$

For Ge, $\lambda_c = \dfrac{1.24}{0.67} = 1.85\ \mu m$

Thus it can be seen that Si-based photodiodes can be used only for the first window whereas Ge-based photodiodes can be used for the second and third windows.

The P-I-N detector is an expression of p–i–n detector, and is specifically developed for optical communication applications. In an ordinary junction photodiode, the 'active' region where the photons are 'captured' is very limited. Besides, the reverse bias voltage applied across the device generates an electric field which is confined to the depletion region. Thus the photons absorbed by the device outside this depletion region are 'lost' and contribute to the low quantum efficiency of this device. The situation can be drastically improved by introducing, between the p and n regions, a large intrinsic (i) region as shown in Fig. 7.3. This detector can now be called the P-I-N detector. The external bias voltage applied to this device will extend to the entire extended depletion region as shown in the figure. Besides, the large depletion region will enhance the capture area for the photons. Additionally, the presence of an electric field throughout this depletion region will ensure the speedy movement of charged carriers, thus reducing transit time effects and making the detector 'fast'. The V–I characteristics of the P-I-N diode for different bias voltages are shown in Fig. 7.4. From this, we can see that by properly reverse biasing the device, a near perfect linear relationship between the optical input power and the electrical output current of the detector can be obtained.

Example 7.3 On a photodiode operating at the wavelength 1.3 μm, 3×10^{11} photons are incident, which results in the generation of 1.5×10^{11} hole–electron pairs. Calculate the quantum efficiency and responsivity of such a photodiode.

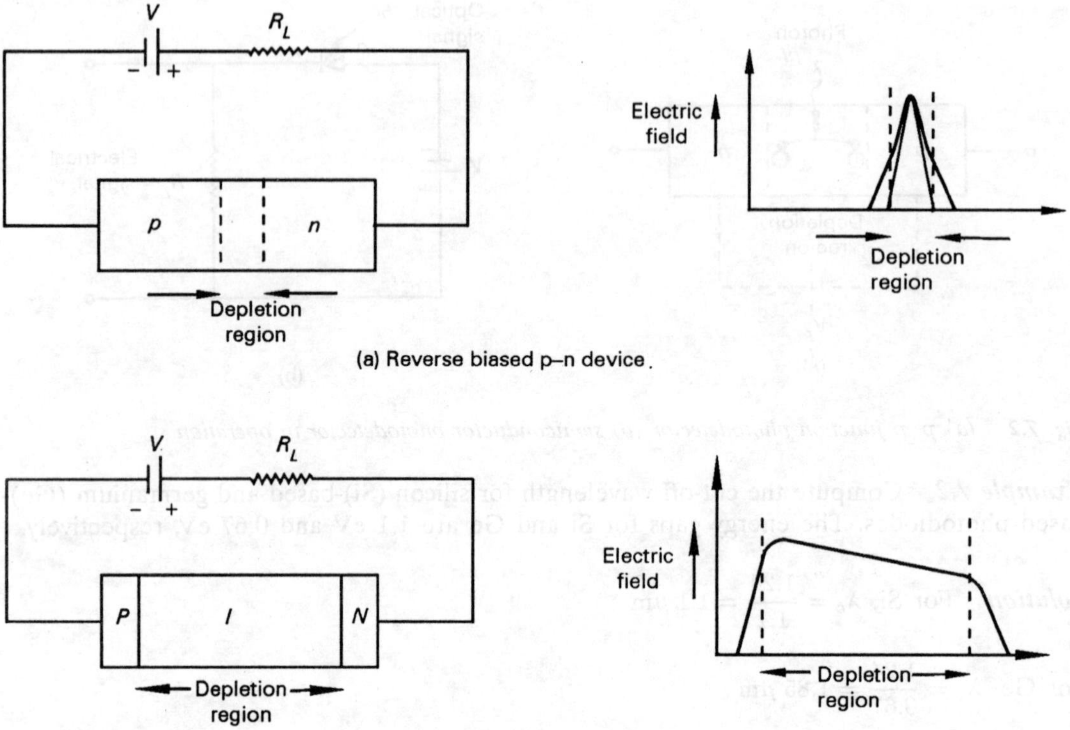

(a) Reverse biased p–n device.

(b) Reverse biased P-I-N device

Fig. 7.3 Electric field distribution in photodetectors. (a) Reverse biased p–n device (b) Reverse biased P-I-N device

Solution As per the definition of quantum efficiency

$$\eta = \frac{\text{no. of hole-electron pairs generated}}{\text{no. of incident photons}}$$

$$= \frac{1.5 \times 10^{11}}{3 \times 10^{11}} = 0.5$$

The responsivity of a photodetector is defined as the ratio of the photocurrent I_p to the incident optical power P_o

$$R = \frac{I_p}{P_o}$$

The number of photons contained in the incident optical beam of power P_o can be expressed as P_o/hf, where f is the frequency of operation ($f = c/\lambda$). Thus, I_p can be expressed as

$$I_p = \frac{\eta P_o e}{hf}$$

$$\therefore \quad R = \frac{I_p}{P_o} = \frac{\eta e}{hf} = \frac{\eta e \lambda}{hc}$$

$$= \frac{0.5 \times 1.602 \times 10^{-19} \times 1.3 \times 10^{-6}}{6.626 \times 10^{-34} \times 2.998 \times 10^{8}}$$

$$= 0.6 \text{ amp/watt.}$$

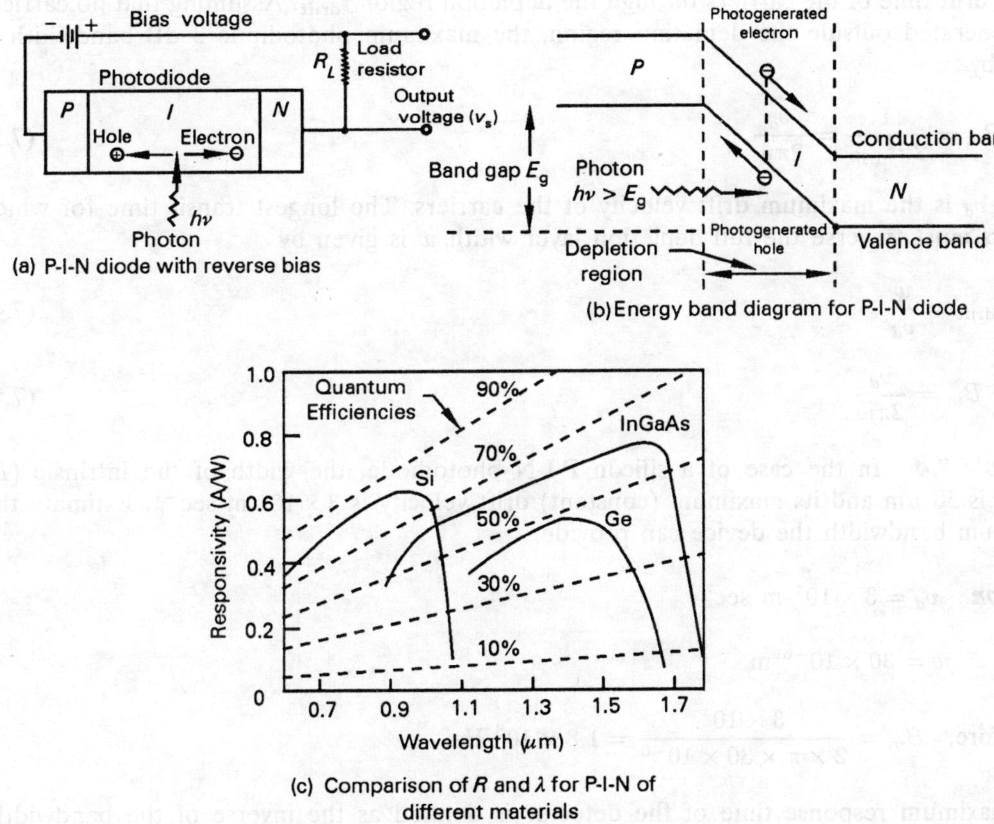

Fig. 7.4 Characteristics and operation of a P-I-N diode (a) P-I-N diode with reverse bias (b) energy band diagram for P-I-N diode (c) comparison of R and λ for P-I-N of different materials.

The responsivity of a photodetector is a very important parameter which in conjunction with the dark current gives an idea about the minimum detectable signal (mds) of a photodetector. Further, it is possible to estimate the signal current from this parameter which, in turn, helps in designing the subsequent stages of the optical receiver. The responsivity is a function of wavelength and quantum efficiency. The quantum efficiencies of Si, Ge and InGaAs are compared in Fig. 7.4. From this, it can be seen that though the Ge P-I-N detectors cover the optical windows at 1.3 μm and 1.55 μm, the InGaAs detector exhibits higher quantum efficiencies at these wavelengths. Thus, though Ge detectors are less expensive, InGaAs detectors are preferred in long-haul optical communication links where the optical signal levels to be detected are very low. Besides, Ge detectors are more noisy with larger dark currents and thus, the minimum detectable optical signals are relatively higher.

The speed of response or the response time of the photodetector is very crucial in determining the information-carrying capability of the optical link. In digital links, this is determined by the rise-time of the photodetector. Although there are other factors which affect the response time of the photodiode, the ultimate bandwidth of the device is limited by the drift time of the carriers through the depletion region t_{drift}. Assuming that no carriers are generated outside the depletion region, the maximum photodiode 3 dB bandwidth is given by

$$B_m = \frac{1}{2\pi t_{\text{drift}}} = \frac{v_d}{2\pi w} \tag{7.3}$$

where v_d is the maximum drift velocity of the carriers. The longest transit time for which carriers must traverse the full depletion layer width w is given by

$$t_{\text{drift}} = \frac{w}{v_d} \tag{7.4}$$

$$\text{Thus,}\quad B_m = \frac{v_d}{2\pi w} \tag{7.5}$$

Example 7.4 In the case of a silicon P-I-N photodiode, the width of the intrinsic (I) region is 30 μm and its maximum (constant) drift velocity is 3×10^4 m sec^{-1}. Estimate the maximum bandwidth the device can provide.

Solution $v_d = 3 \times 10^4$ m sec^{-1}

$w = 30 \times 10^{-6}$ m

$$\text{Therefore,}\quad B_m = \frac{3 \times 10^4}{2 \times \pi \times 30 \times 10^{-6}} = 1.8 \times 10^8 \text{ Hz.}$$

The maximum response time of the detector is defined as the inverse of the bandwidth, and is given by

$1/B_m = 5.4$ nsec.

Besides the drift time of the carriers, another factor which affects the detector response time is the capacitance of the photodiode. A reverse-biased photodiode exhibits a voltage-dependent capacitance caused by the stored charge at the junction. The junction capacitance C_j is given by

$$C_j = \frac{\epsilon_s A}{w} \tag{7.6}$$

where ϵ_s is the permittivity of the semiconductor material and A is the diode junction area. If the width of the depletion layer is small it will increase the capacitance. To this, the capacitance effects due to the leads and packaging of the device must also be added. The overall capacitance of the device must be minimized in order to reduce the RC time constant which affects the detector response time. The effects due to the diffusion time of the carriers on the response time of the detector can be ignored because most of the carriers are generated within the depletion region.

Another important photodetector in optical communication is the Avalanche Photodetector (APD). This is analogous to the photomultiplier tube (PMT) with an internal gain. It can also be referred to as the semiconductor equivalent of PMT where a large electric field will accelerate the primary charge carrier to cause avalanche breakdown. The structure of an APD and the electric field distribution inside it due to a large biasing voltage are shown in Fig. 7.5. To create a larger electric field for causing internal gain, we require high biasing voltages of the order of 300 to 400 volts. The large electric field will cause the electrons and holes to accelerate and create additional electron–hole pairs due to collisions. Carrier multiplication factors as large as 10^4 can be obtained by proper selection of materials and suitable engineering of the device. The disadvantages of the APD are that it demands high biasing voltage and it is also more noisy than the P-I-N detector. As in the case of P-I-N diode, the response time of an APD is limited by the charge carrier transit time and the RC time constant. For long-haul optical communication links where the received optical signals are very weak, APD is preferred over a P-I-N detector.

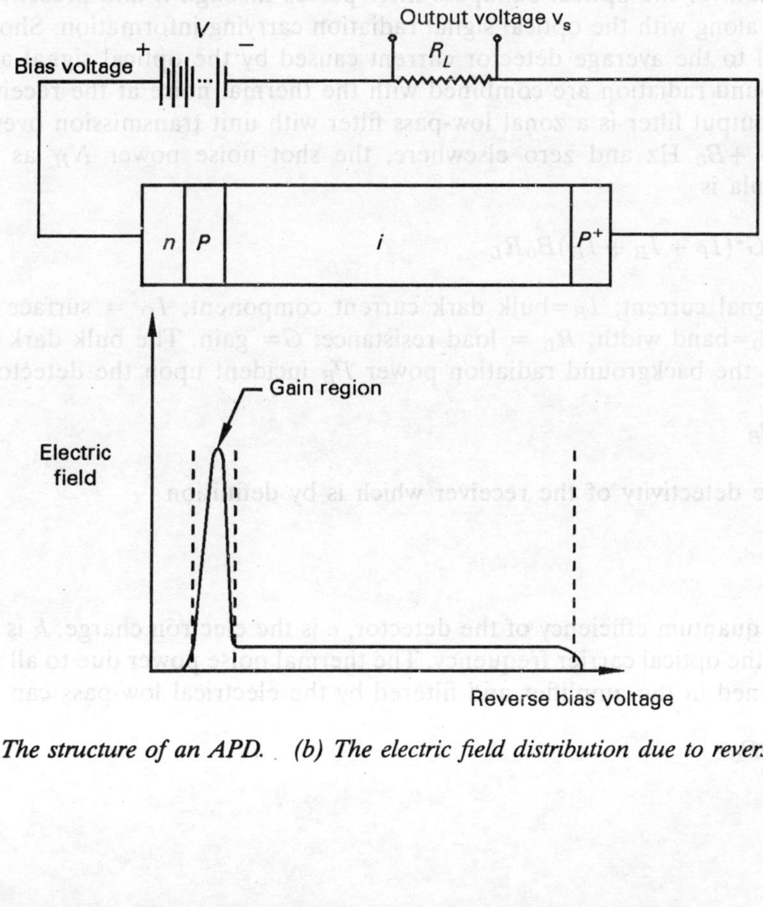

Fig. 7.5 *(a) The structure of an APD. (b) The electric field distribution due to reverse bias voltage.*

7.5 Optical receivers

Optical receivers are the final stage in an optical communication system from which the information sent from the transmitting end is obtained. The block diagram of a typical baseband analog optical receiver is shown in Fig. 7.6.

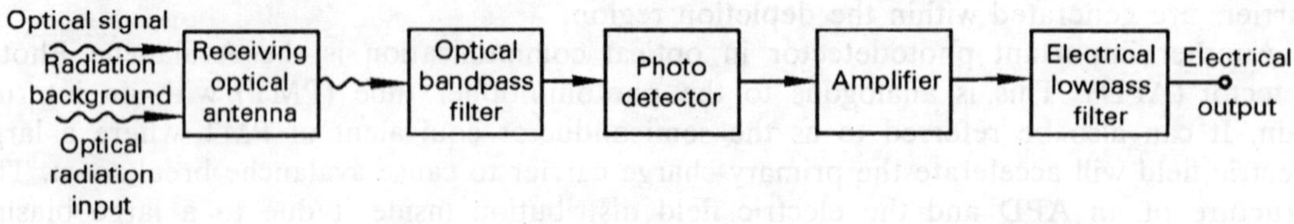

Fig. 7.6 Baseband analog direct detection receiver.

The photodetector is the heart of the optical receiver and is crucial to the overall performance of the receiver. Along with the signal radiation, the background optical radiation is also gathered by the receiving optical antenna and is injected into the photodetector after passing through an optical bandpass filter with its central frequency corresponding to the optical signal carrier frequency f_c. Even though the optical bandpass filter substantially reduces the background optical radiation, some background radiation corresponding to the optical bandwidth of the optical bandpass filter passes through it and presents itself to the photodetector along with the optical signal radiation carrying information. Shot noise which is proportional to the average detector current caused by the optical signal and noise due to the background radiation are combined with the thermal noise at the receiver output. If the electrical output filter is a zonal low-pass filter with unit transmission over a frequency range $-B_0$ to $+B_0$ Hz and zero elsewhere, the shot noise power N_H as given by the Schottky formula is

$$N_H = 2eG^2(I_P + I_B + I_D)B_0 R_L \tag{7.7}$$

where I_P= signal current; I_B=bulk dark current component; I_D = surface dark current component; B_0=band width; R_L = load resistance; G= gain. The bulk dark current component due to the background radiation power P_B incident upon the detector is given by

$$I_B = DP_B \tag{7.8}$$

where D is the detectivity of the receiver which is by definition

$$D = \frac{\eta e}{h f_c} \tag{7.9}$$

where η is the quantum efficiency of the detector, e is the electron charge, h is Planck's constant and f_c is the optical carrier frequency. The thermal noise power due to all resistive components combined in the amplifier and filtered by the electrical low-pass can be written as

$$N_T = kTB_0 \tag{7.10}$$

If i_p is the instantaneous photodetector current due to the optical signal impinging on it, then the instantaneous receiver output signal voltage v_s is equal to

$$v_s = i_p R_L = GDP_C R_L \tag{7.11}$$

where P_C is the optical carrier power. The average signal power at the output of the optical receiver is defined as the time average of v_s^2 over a time period comparable to the reciprocal of the output electrical filter bandwidth $-B_0$. Thus, for the demodulated optical carrier, the signal power can be written as

$$S = \frac{v_s^2}{R_C} = (GDP_C)^2 R_L \tag{7.12}$$

This is a measure of the optical signal power at its peak value since any intensity modulation reduces the carrier intensity. Since any photodetector basically responds only to changes in the intensity of the optical signal falling on it, any other type of modulation should be first converted to intensity modulation before it impinges on the photodetector.

A common performance criterion of any receiver in a communication system is the signal-to-noise ratio (SNR) which is defined as the ratio of the average signal power to the average noise power at some nodal point in the receiver chain. It is important to note that in optical communication the SNR is a function of the carrier power as can be seen from the following equation

$$\frac{S}{N} = \frac{\left[\dfrac{G\eta e}{hf_c}\right]^2 R_L P_C^2}{2eB_0G^2\left[\dfrac{\eta e}{hf_c}[P_C + P_B] + I_D\right] R_L + 4kTB_0} \tag{7.13}$$

If the photodetector internal gain (G) is unity as is the case in P-I-N photodetector, the thermal noise is much greater than the shot noise and the SNR becomes

$$\frac{S}{N} = \left[\frac{\eta e}{hf_c}\right]^2 \frac{R_L P_C^2}{4kTB_0} \tag{7.14}$$

If, on the other hand, the photodetector has a large internal gain ($G \gg 1$), the shot noise given by the first term in the denominator of Eq.(7.13) predominates over the thermal noise and the SNR reduces to

$$\frac{S}{N} = \left[\frac{\eta e}{hf_c}\right]^2 \left[\frac{e}{2B_0}\right] \left[\frac{P_C^2}{\dfrac{\eta e}{hf_c}[P_C + P_B] + I_D}\right] \tag{7.15}$$

The optical receiver discussed so far is referred to as the Intensity-modulation/Direct-detection (IM/DD) receiver which normally has poor sensitivity. This system is the optical equivalent of the crystal radio receiver used in the early days of radio communication. After the introduction of the super heterodyne receiver has revolutionized radio communication, it is tempting to extend these ideas to optical communication. Coherent optical communication systems, based on the heterodyne principle, have two fundamental advantages over the IM/DD systems. The first advantage is the improved sensitivity of the optical receiver by over 20 dB which makes long-haul optical communication systems possible. Secondly, the significant improvement in the receiver selectivity makes multichannel optical communication systems feasible.

Example 7.5 Suppose that in an optical communication system P-I-N detector the internal gain G is almost unity. Assuming the thermal work limited case, calculate the signal-to-noise ratio (SNR) for an optical signal of 0.4 μW falling on the detector whose responsivity D is 0.5 A/W. The load resistance is 50 Ω, the receiver bandwidth B_0 is 10 MHz and its temperature is 300 K.

Solution The electrical signal power across the load resistance of the photodiode is given by

$$S = (GDP_C)^2 R_L$$

Here, $G = 1$, $D = 0.5$ A/W; $P_C = 0.4 \times 10^{-6}$ W and $R_L = 50\,\Omega$

Therefore, $S = -2 \times 10^{-12}$ W

In the thermal noise-limited case, the noise power is given by the equation

$$N = 4kTB_0$$

Now, $k = 1.38 \times 10^{-23}$ J/K, $T = 300$ K and $B_0 = 10 \times 10^6$ Hz

Thus, $N = 1.66 \times 10^{-13}$ W.

So the signal-noise ratio (SNR) is

$$\frac{S}{N} = \frac{2 \times 10^{-12}}{1.66 \times 10^{-13}} = 10.8 \text{ dB}$$

The SNR value can be improved by cooling the detector and thereby decreasing the value of T.

7.6 Digital optical receivers

Because of the obvious advantages of digital communication systems over analog systems, the information is often sent in the form of bits of 0s and 1s. The number of bits that can be transmitted and received unambiguously will determine channel capacity with units as 'bits/second'. The signals in a digital communication system are significantly perturbed by noise which causes the inter-symbol interference (ISI). This creates a confusion between

0s and 1s, thus leading to bit-error-rate (BER). The ISI in an optical receiver is caused by various capacitive components in the circuitry and the electrical filters used. A typical detected signal with and without ISI is shown in Fig. 7.7.

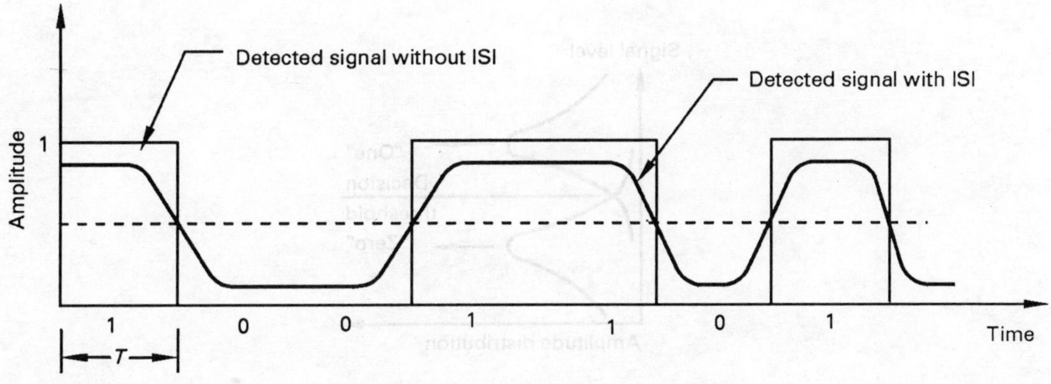

Fig. 7.7 *Detected optical signal with and without ISI.*

One obvious way to decrease the influence of ISI is to increase the available bandwidth by properly choosing the components and suitably designing the filters. In case of infinite bandwidth, the detected signal will retain its original shape. However, large bandwidth will open the flood gates for noise which will increase the BER. Thus, there is an obvious trade-off between bandwidth and transmission quality.

A typical optical receiver for digital communication links is shown in Fig. 7.8. The basic digital signal formats namely the return-to-zero (RZ) and non-return-to-zero (NRZ) and the BER concept based on confusion between 1s and 0s due to noise are shown in Fig. 7.9. From the figure, we can see that it is possible to make an error in judging 1 and 0 due to noise and inter-symbol interference. This is caused by making the threshold between 'zero' and 'one' hazy, and thus causing a bit error. According to the specifications laid down by the International Telecommunication Union (ITU), for a reliable communication link, it is recommended that one should not commit more than one mistake in determining the bits in a sequence of 10^9 bits.

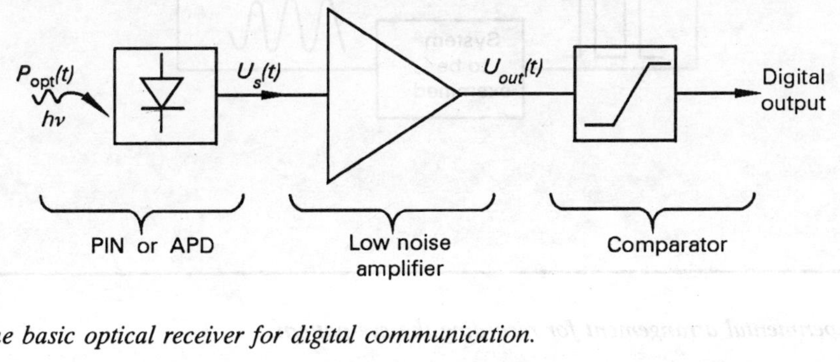

Fig. 7.8 *The basic optical receiver for digital communication.*

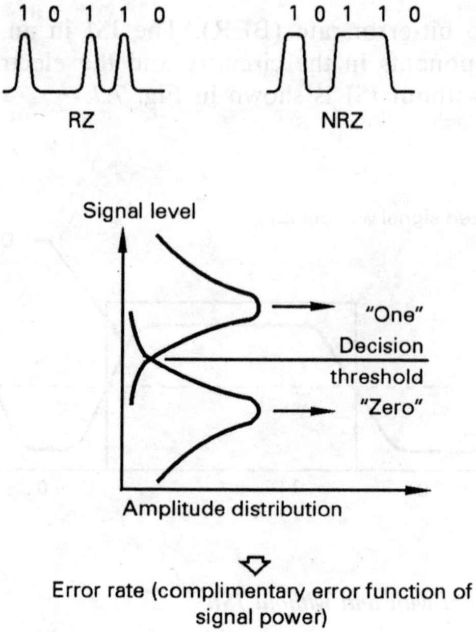

Fig. 7.9 Digital signal transmission formats.

In the development and realization of reliable digital optical communication systems, we have to determine the degree of BER required. For this purpose, 'eye-pattern' is recorded using an experimental arrangement as shown in Fig. 7.10.

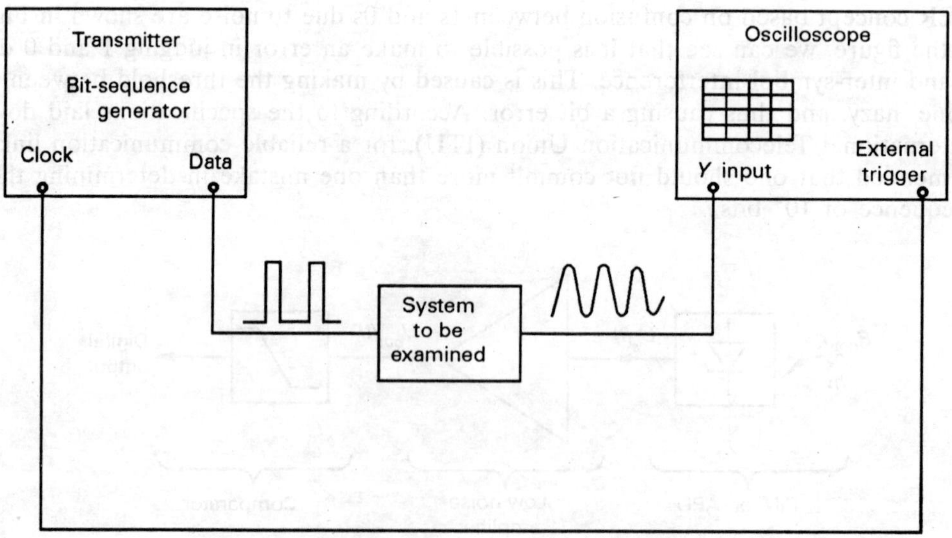

Fig. 7.10 Experimental arrangement for recording the eye-pattern.

A typical eye-pattern recorded due to a sequence of bits at the receiver is shown in Fig. 7.11. As long as the 'eye' is open, we can say that the digital link is reliable but when the 'eye' closes, the link is unreliable. This is a simple and effective way of monitoring the health of a digital optical communication link. Of course, there are more reliable but expensive ways to measure BER, which are rather difficult to operate in the field.

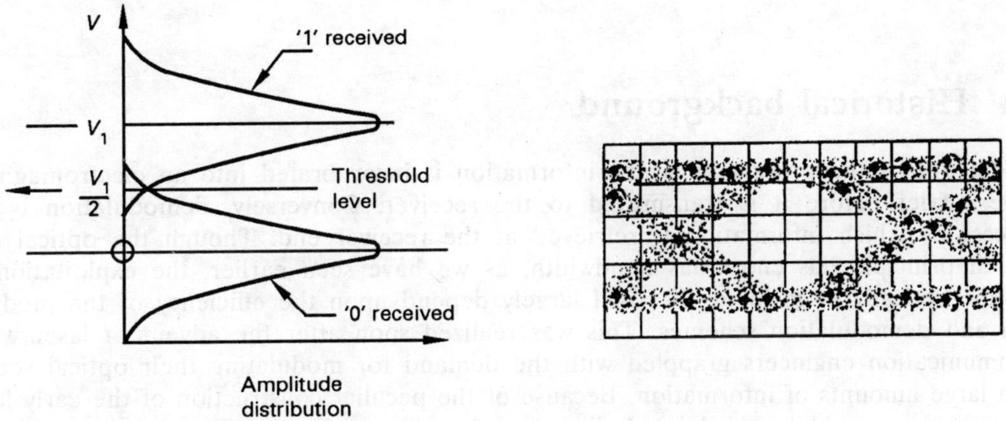

Fig. 7.11 *Typical eye-pattern observed.*

8 Modulation and demodulation schemes in optical communication

8.1 Historical background

Modulation is the process by which information is incorporated into an electromagnetic wave carrier before it is transmitted to the receiver. Conversely, demodulation is the process by which information is retrieved at the receiver end. Though the optical carrier in principle has enormous bandwidth, as we have seen earlier, the exploitation of the theoretically large bandwidth will largely depend upon the efficiency of the modulation and demodulation schemes. This was realized soon after the advent of laser when communication engineers grappled with the demand for modulating their optical source with large amounts of information. Because of the peculiar construction of the early laser systems, they could be modulated directly only with small amounts of information. In order to improve the situation, scientists were forced to think in terms of external modulation which added to the complexity and cost of the system. The appearance of the laser diode on the scene made direct modulation feasible at high-bit rates of information. Whatever be the modulation scheme, the photodetector can, in principle, respond only to changes in the optical intensity. This, however, limits the possibilities for demodulation schemes. But over the years, communication engineers have mastered the art of harnessing the large amounts of information capacity inherently available in an optical carrier by developing appropriate modulation and demodulation schemes.

Table 8.1 *Optical modulation methods*

	Analog	Pulse	Digital
Information signal	Continuous in time	Sampled in time	Sampled in time
Optical carrier parameter	Continuous	Quantized	Quantized and coded
Examples	Analog intensity modulation (IM)	Pulse intensity modulation (PIM)	Pulse code modulation (PCM)

8.2 Fundamentals of modulation schemes

Modulation of the optical carriers, i.e., the electromagnetic waves at optical frequencies, is different from those at radio frequencies in more than one way. This arises out of the basic differences between the devices which generate electromagnetic waves at radio and optical frequencies. At optical frequencies, modulators are designed to directly operate on the carrier intensity rather than on the amplitude of the electric field as is the case with radio frequency modulation. In principle, the modulation of an optical carrier can be effected on any of its parameters namely the amplitude, intensity, frequency, phase and polarization. A good classification of optical modulation methods is given in Table 8.1. Figure 8.1 illustrates the three examples of analog intensity modulation, pulse intensity modulation and pulse code modulation.

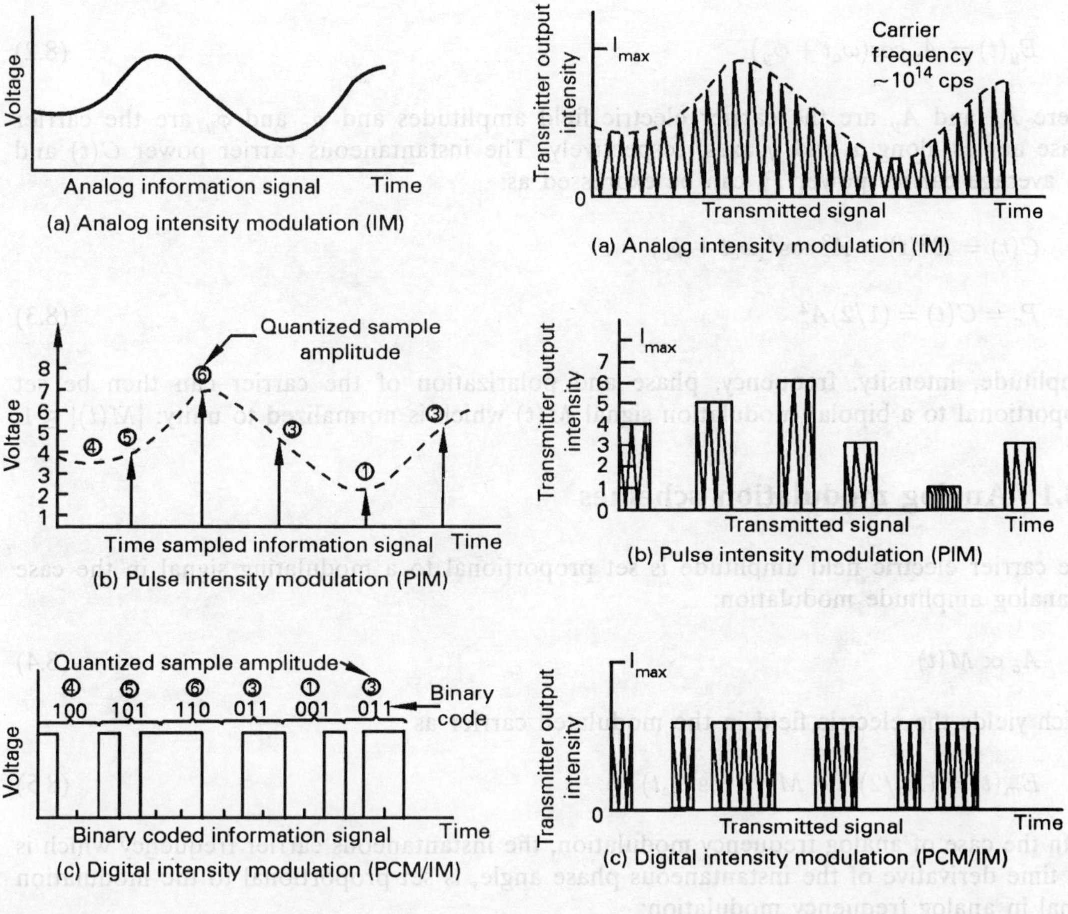

Fig. 8.1 Analog modulated waveforms (a) Analog intensity modulation (IM) (b) Pulse intensity modulation (PIM) (c) Pulse code modulation (PCM).

8.3 Modulation spectral analysis

The instantaneous electric field $E(t)$ of an unmodulated optical carrier may be expressed as

$$E(t) = A_c \cos(\omega_c t + \phi_c) \tag{8.1}$$

where A_c is the amplitude of the carrier electric field, $\omega_c/2\pi = f_c$ is the frequency of the carrier, and ϕ_c is the phase angle of the carrier.

The electric field $E(t)$ which is normally confined to the xy plane in a carrier propagating along the z-direction can be expressed as

$$E_x(t) = A_x \cos(\omega_c t + \phi_x)$$

$$E_y(t) = A_y \cos(\omega_c t + \phi_y) \tag{8.2}$$

where A_x and A_y are the carrier electric field amplitudes and ϕ_x and ϕ_y are the carrier phase angles along x and y axes, respectively. The instantaneous carrier power $C(t)$ and the average carrier power P_c can be expressed as:

$$C(t) = E^2(t) = A_c^2 \cos^2(\omega_c t + \phi_c)$$

$$P_c = C(t) = (1/2)A_c^2 \tag{8.3}$$

Amplitude, intensity, frequency, phase and polarization of the carrier can then be set proportional to a bipolar modulation signal $M(t)$ which is normalized to unity: $|M(t)| \leq 1$.

8.3.1 Analog modulation schemes

The carrier electric field amplitude is set proportional to a modulating signal in the case of analog amplitude modulation:

$$A_c \propto M(t) \tag{8.4}$$

which yields the electric field in the modulated carrier as

$$E_m(t) = (A_c/2)[1 + M(t)] \cos(\omega_c t) \tag{8.5}$$

In the case of analog frequency modulation, the instantaneous carrier frequency which is the time derivative of the instantaneous phase angle, is set proportional to the modulation signal in analog frequency modulation:

$$\frac{d\phi_c}{dt} \propto M(t) = K_F M(t) \tag{8.6}$$

where K_F=FM modulation index. This results in the electric field of the modulated carrier as

$$E_m(t) = A_c \cos\left[\omega_c t + K_F \int M(t)dt\right] \tag{8.7}$$

where $\omega_d/2\pi$ is the peak frequency deviation from the carrier frequency f_c.

For phase modulation, the instantaneous carrier phase ϕ_c is set proportional to the modulation signal

$$\phi_c \propto M(t) \tag{8.8}$$

This results in

$$E_m(t) = A_c \cos[\omega_c t + \phi_k M(t)] \tag{8.9}$$

where ϕ_k is the phase deviation constant. For analog intensity modulation, the square of the carrier electric field amplitude is set proportional to the modulation signal

$$A_c^2 \propto M(t) \tag{8.10}$$

The resulting modulated carrier intensity is given by

$$C_m(t) = A_c^2/2[1 + M(t)] \cos^2(\omega_c t) \tag{8.11}$$

The spectral analysis of these different modulation schemes can be carried out on the same lines as in the case of a general electromagnetic wave modulation.

8.3.2 Pulse modulation scheme

In the case of pulse modulation, there are certain additional schemes which are possible when compared to the analog modulation. Besides, there are some well-known advantages of the pulse modulation schemes like power efficiency in implementing them. In the case of pulse amplitude modulation (PAM), the waveform of the modulated wave amplitude can be written as

$$E_m(t) = A_c/2[1 + M(t_n)] \cos(\omega_c t) \quad \text{for } t_n \leq t \leq t_n + \tau \tag{8.12}$$

where $M(t_n)$ is the amplitude of the information (which may be quantized or continuous) of the sample time t_n, and τ is the constant carrier pulse duration. In the case of pulse frequency modulation (PFM), the modulated carrier electric field can be represented as

$$E_m(t) = A_c \cos\left[\omega_c t + \omega_d \int M(t_n) dt\right] \quad \text{for } t_n \leq t \leq t_n + \tau \tag{8.13}$$

Since $M(t_n)$ is constant for a particular data sample, the carrier pulse frequency is simply shifted by some fraction of the maximum frequency deviation for the carrier pulse duration. For pulse intensity modulation (PIM), the carrier intensity may be represented as

$$C_M(t) = \frac{A_c^2}{2}[1 + M(t_n)] \cos^2 \omega_c t \quad \text{for } t_n \leq t \leq t_n + \tau \tag{8.14}$$

In pulse modulation scheme, it is possible to consider pulse duration as a possible modulation parameter. The carrier electric field in the pulse duration modulation (PDM) can be represented as

$$E_m(t) = A_c \cos(\omega_c t) \quad \text{for } t_n \leq t \leq t_n + \tau_\omega \tag{8.15}$$

where the time duration τ_ω of the carrier pulse is given by

$$\tau_\omega = \tau_p/2[1 + M(t_n)]$$

The maximum pulse duration is limited to less than the sample period τ_p to enable the receiver to detect the carrier pulse leading edge for synchronization purposes. Pulse width modulation (PWM) is a variant of PDM where the pulse duration is made to vary symmetrically about the data sample time t_n.

In pulse position modulation (PPM), the carrier electric field of duration τ can be expressed as

$$E_m(t) = A_c \cos(\omega_c t) \quad \text{for } t_n + \tau_d \leq t \leq t_n + \tau_d + \tau \tag{8.16}$$

where the time delay τ_d of the leading edge of the carrier pulse with respect to the sample time t_n is given by

$$\tau_d = \tau_p/2[1 + M(t_n)] \tag{8.17}$$

In order to prevent pulse overlap in time-adjacent sample periods, the maximum carrier pulse delay should always be less than the sample period τ_p. A marker pulse is generated and transmitted in PPM systems in order to provide for synchronization. Pulse rate modulation (PRM) system can either be continuous or time sampled in which the instantaneous rate of generation of carrier pulses is made proportional to the information signal amplitude.

8.3.3 Digital modulation schemes

The digital modulation methods will involve sampling/coding processes. They are immune to channel noise and are generally considered to be superior to the earlier modulation schemes. At optical frequencies, pulse code modulation/intensity modulation (PCM/IM) is also referred to as PCM/ASK, i.e., amplitude shift keying, where the carrier intensity is set at maximum to represent a 'one' bit or at a minimum to represent a 'zero' bit of a binary code.

In PCM-frequency modulation (PCM/FM) which is also referred to as PCM-frequency-shift keying (PCM/FSK), the carrier frequency is set at one of the two possible values to represent either a 'one' or a 'zero' bit of binary code of information signal sample amplitude. In a similar way, for PCM-phase modulation, the carrier phase angle is set at a phase angle of zero or π radians in respect to a phase reference to represent a 'one' bit or a 'zero' bit of binary code of information signal amplitude. In the case of PCM-polarization modulation (PCM/PM), the carrier is set in vertical polarization to represent a 'one' bit and horizontal polarization to represent a 'zero' bit of binary code of information signal sample amplitude.

The modulation of a sinusoidal optical carrier by digital signals leads to amplitude shift keying (ASK) or ON–OFF keyed (OOK) signals, frequency shift keyed (FSK) signals or phase shift keyed (PSK) signals. These signals have bandpass spectra. Several modifications to these standard formats are also possible. The pictorial representation of these schemes is given in Fig. 8.2. It is not possible to give a pictorial representation of PCM/PL as in other cases. Of all the above-mentioned digital schemes, PSK signals offer the best noise performance, i.e., for a given signal-to-noise ratio (SNR), the bit-error-rate (BER) is the smallest. Though it is rather difficult to visualize the PCM/PL scheme, it is a very attractive proposition at optical frequencies where the state of polarization can be organized and detected with relative ease. Since the photodetector can decipher only intensity changes, all the modulation schemes have to be ultimately converted into intensity changes before they arrive at the photodetector for demodulation.

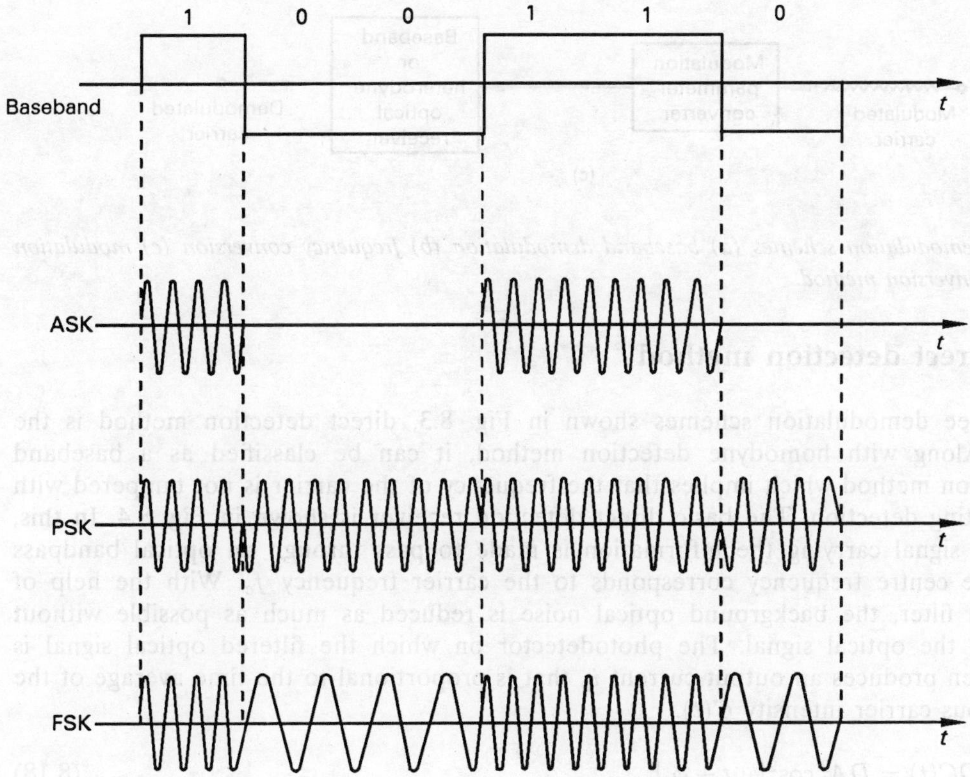

Fig. 8.2 Digital modulated waveforms (a) Amplitude shift keyed signal (ASK) (b) Frequency shift keyed signal (FSK)(c) Phase shift keyed signal (PSK).

8.4 Demodulation methods

Optical demodulation or detection is the process by which information is extracted from the carrier after it arrives at the receiving end. The demodulation schemes can be broadly classified into three categories as shown in Fig. 8.3. They are: (i) direct detection method, (ii) heterodyne detection method and (iii) homodyne detection method.

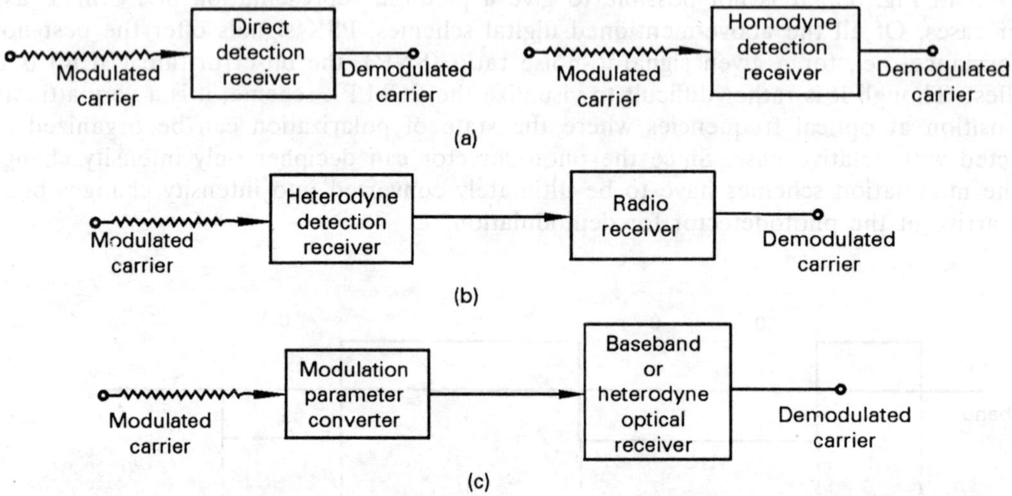

Fig. 8.3 Demodulation schemes (a) baseband demodulation (b) frequency conversion (c) modulation parameter conversion method.

8.4.1 Direct detection method

Of the three demodulation schemes shown in Fig. 8.3, direct detection method is the simplest. Along with homodyne detection method, it can be classified as a baseband demodulation method which implies that the frequency of the carrier is not tampered with while effecting detection. The basic direct detection receiver is shown in Fig. 8.4. In this, the optical signal carrying the information is made to pass through an optical bandpass filter whose centre frequency corresponds to the carrier frequency f_c. With the help of this optical filter, the background optical noise is reduced as much as possible without attenuating the optical signal. The photodetector on which the filtered optical signal is incident then produces an output current i_p that is proportional to the time average of the instantaneous carrier intensity $C(t)$.

$$i_p = DC(t) = DA_c^2 \cos^2[\omega_c t + \phi_c] \tag{8.18}$$

where D is the detectivity of the photodetector (also referred to, sometimes, as responsitivity R or as the conversion factor). Its unit is amp/watt (A/W). The time averaging process of the photodetector on the carrier intensity makes the output photocurrent i_p insensitive to changes in frequency, phase and polarization of the carrier.

$$i_p = DA_c^2/2 \tag{8.19}$$

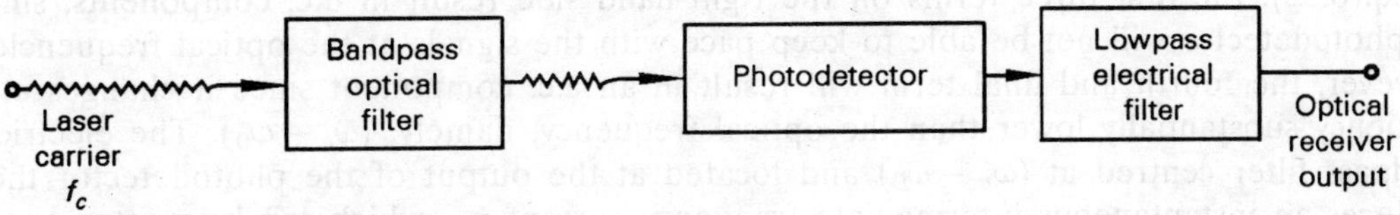

Fig. 8.4 *The basic direct detection receiver.*

As such the direct detection receiver can technically be used only for amplitude modulation (AM) and intensity modulation (IM). However, since AM cannot be affected by an optical transmitter because of the way in which the electromagnetic waves are generated and transmitted at these frequencies. Besides, only intensity modulation (IM) can be demodulated by a direct detection receiver.

8.4.2 Heterodyne detection receiver

Optical heterodyne detection receiver is the equivalent of a super-heterodyne receiver at radio frequencies where there is a local oscillator incorporated in the detection process. The block diagram of an optical heterodyne receiver is shown in Fig. 8.5 where the local laser oscillator output is mixed with the incoming signal with the help of a beam splitter. The electric fields of the carrier and local oscillator may be expressed as

$$E(t) = A_c \cos[\omega_c t + \phi_c]$$

$$\text{and} \quad L(t) = A_0 \cos[\omega_0 t + \phi_0] \tag{8.20}$$

When the carrier and local oscillator beams are aligned perpendicular to the photodetector surface in close proximity, the resulting photodetector output current is given by

$$i_p = D[E(t) + L(t)]^2 \tag{8.21}$$

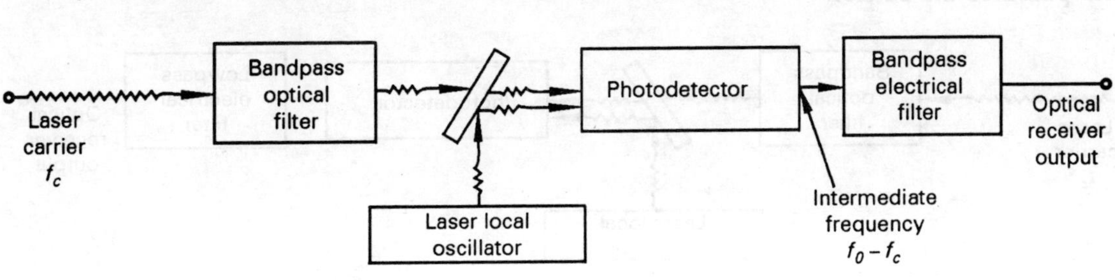

Fig. 8.5 *The heterodyne detection receiver.*

which, when combined with Eq.(8.20), yields

$$i_p = D\{A_c^2 \cos^2[\omega_c t + c] + A_0^2 \cos^2[\omega_0 t + \phi_0] + A_c A_0 \cos[(\omega_0 + \omega_c)t + (\phi_c + \phi_0)] + A_c A_0 \cos[(\omega_c - \omega_0)t + (\phi_c - \phi_0)]\} \quad (8.22)$$

In Eq.(8.22), the first three terms on the right-hand side result in d.c. components, since the photodetector will not be able to keep pace with the signals at the optical frequencies. However, the fourth and final term will result in an a.c. component since it changes at a frequency substantially lower than the optical frequency, namely, $(\omega_c - \omega_0)$. The electrical bandpass filter centred at $(\omega_c - \omega_0)$ and located at the output of the photodetector then produces an instantaneous intermediate frequency current i_{IF} which can be written as

$$i_{IF} = DA_c A_0 \cos[(\omega_c - \omega_0)t + (\phi_c - \phi_0)] \quad (8.23)$$

From this, it can be seen that the output of the heterodyne detection receiver is sensitive to changes in the phase of the carrier.

8.4.3 Homodyne detection optical receiver

Since it is not always possible to tune the laser oscillator as easily as the radio frequency oscillator, it is often difficult to realize the heterodyne optical receiver in practice. However, it is always possible to use the same laser, which is used as the transmitter, as the local oscillator with much smaller power output (Fig. 8.6). In such a case, the receiver output current reduces to

$$i_{IF} = A_c A_0 \cos(\phi_0 - \phi_c) \quad (8.24)$$

since $\omega_0 = \omega_c$. The phase of the local oscillator in the optical domain can be set at a desired value easily by a phase plate. Hence, if we make $\phi_0 = \pi/2$, Eq.(8.24) can be expressed as

$$i_{IF} = A_c A_0 \cos(\pi/2 - \phi_c) = A_c A_0 \sin \phi_c \quad (8.25)$$

For small changes in the phase angle ϕ_c, Eq.(8.25) becomes

$$i_{IF} \approx A_c A_0 \phi_c \quad (8.26)$$

Thus, the output of the homodyne optical receiver can detect changes in the amplitude and phase of the carrier.

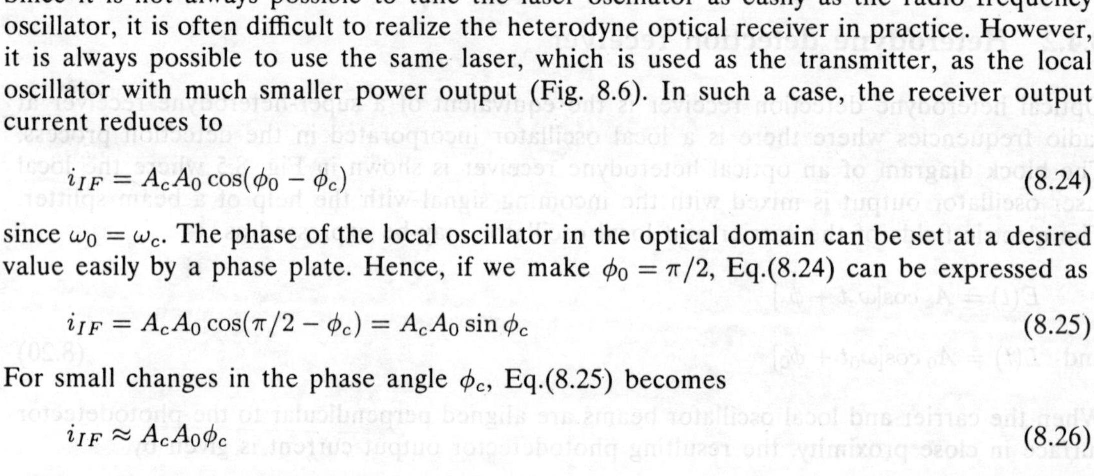

Fig. 8.6 The homodyne detection receiver.

8.4.4 Modulation parameter convertors

Since many of the carrier parameters like frequency and phase offer excellent advantages over intensity modulation, a modulation parameter convertor is used as a front-end to the optical receiver while dealing with such carrier information. The function of the modulation parameter convertor is to convert the FM or PM modulation format into IM format such that the optical receiver can retrieve the information. A typical example of an FM–IM modulation parameter convertor is shown in Fig. 8.7(b) where a skirt filter is placed before the optical receiver. The characteristics of the skirt filter whose attenuation is a function of frequency is shown in Fig. 8.7(a). Skirt filters can be easily realized in practice by the use of well-known optical components.

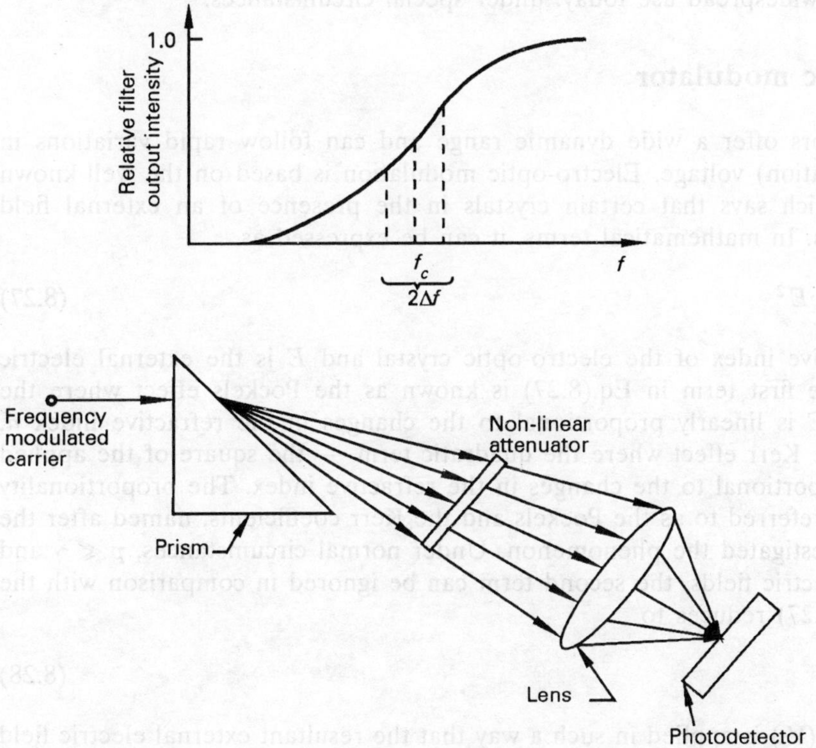

Fig. 8.7 (a) Skirt filter characteristics (b) FM–IM modulation parameter convertor.

8.5 External optical modulators

Since the early laser systems were not easy to modulate directly at high frequencies due to their peculiar constructional features, optical communication researchers were forced to explore the possibility of external optical modulation schemes. Though the He–Ne lasers yielded direct CW modulation through its power supply, it was barely sufficient to demonstrate voice optical communication link in the laboratory. In terms of bandwidth, such a link is like 'scratching the surface' of an inherently large bandwidth which an optical communication link can, and should, provide.

An optical carrier beam can be modulated externally by several methods when the beam is passing through the optical modulator. This will automatically imply that the optical modulator should be transparent to the optical carrier beam. While allowing the optical carrier to pass through it with minimum attenuation, the optical modulator should be able to put information onto the carrier such that a modulated output carrier would emerge from it. Of all the external modulation schemes, electro-optic and acousto-optic modulation methods emerged as the best which could be operated at high frequencies. Even after the advent of semiconductor lasers which are directly modulatable, the miniaturized version of these modulators find widespread use today, under special circumstances.

8.5.1 Electro-optic modulator

Electro-optic modulators offer a wide dynamic range and can follow rapid variations in applied signal (information) voltage. Electro-optic modulation is based on the well-known electro-optic effect which says that certain crystals in the presence of an external field behave as birefringents. In mathematical terms, it can be expressed as

$$\Delta(1/n^2) = \gamma E + pE^2 \tag{8.27}$$

where n is the refractive index of the electro-optic crystal and E is the external electric field applied to it. The first term in Eq.(8.27) is known as the Pockels effect where the applied electric field E is linearly proportional to the changes in the refractive index n. The second term is the Kerr effect where the quadratic term — the square of the applied electric field — is proportional to the changes in the refractive index. The proportionality constants γ and p are referred to as the Pockels and the Kerr coefficients, named after the scientists who first investigated the phenomenon. Under normal circumstances, $p \ll \gamma$ and therefore, for small electric fields, the second term can be ignored in comparison with the first term. Thus, Eq.(8.27) reduces to

$$\Delta(1/n^2) = \gamma E \tag{8.28}$$

The external potential (V) is applied in such a way that the resultant external electric field is oriented along the direction of propagation of optical carrier as shown in Fig. 8.8.

The electric field of the incident electromagnetic wave optical carrier transmitter can be written as

$$E = E_0 \cos \omega t \tag{8.29}$$

This can be represented as a combination of right- and left-circularly polarized components given by

$$E_x = \frac{E_0}{\sqrt{2}} \cos \omega t \quad \text{and} \quad E_y = \frac{E_0}{\sqrt{2}} \cos \omega t \tag{8.30}$$

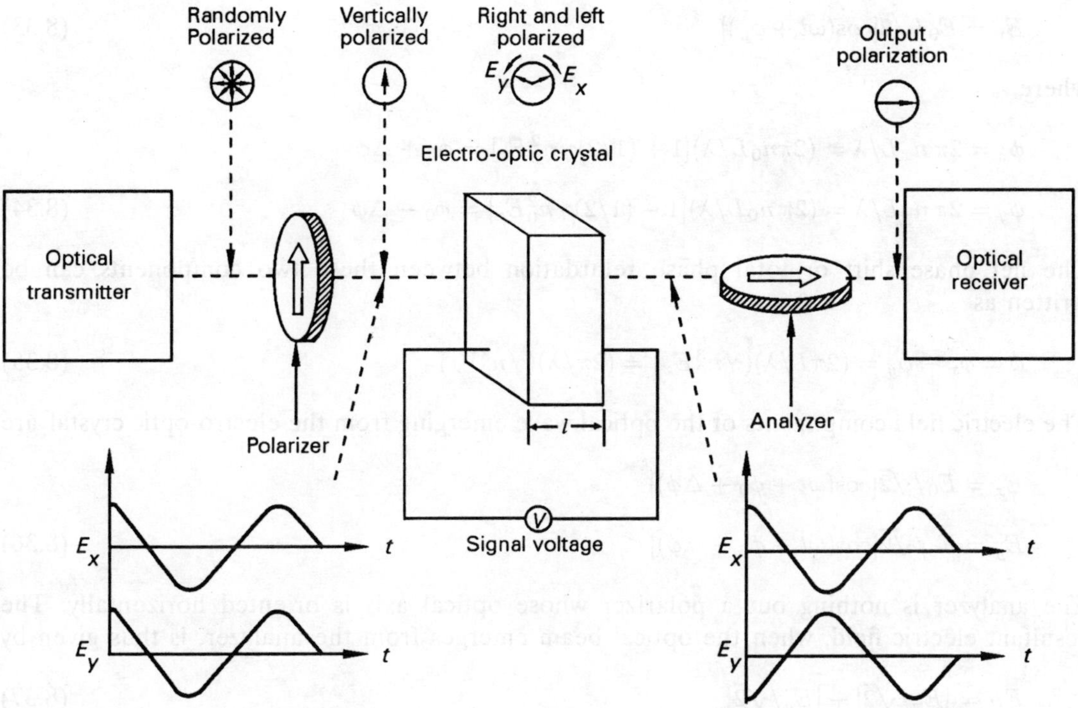

Fig. 8.8 Electro-optic modulator.

Now, when the electric field components (E_x and E_y) enter the electro-optic crystal across which the electric potential (V) is applied, coinciding with the direction of propagation of the optical carrier, they will experience two different refractive indices n_x and n_y which are given by

$$n_x = n_0 + \frac{1}{2}\gamma\, n_0^3 E_z$$

$$n_y = n_0 - \frac{1}{2}\gamma\, n_0^3 E_z \tag{8.31}$$

where $E_z = V/L$, L being the length of the electro-optic crystal. In other words, the velocity phase of propagation of the right-circularly polarized light will be different from that of the left-circularly polarized light. The phase shifts experienced by these components can be expressed as

$$\phi_x = 2\pi\, n_x L/\lambda$$

$$\phi_y = 2\pi\, n_y L/\lambda \tag{8.32}$$

Thus the electric fields of the right- and left-circularly polarized components after they emerge from the electro-optic crystal can be expressed as

$$E_x = E_0/\sqrt{2}[\cos(\omega t + \phi_x)]$$

$$E_y = E_0/\sqrt{2}[\cos(\omega t + \phi_y)] \tag{8.33}$$

where

$$\phi_x = 2\pi\, n_x L/\lambda = (2\pi n_0 L/\lambda)[1 + (1/2)\gamma\, n_0^2 E_z] = \phi_0 + \Delta\phi$$

$$\phi_y = 2\pi\, n_y L/\lambda = (2\pi\, n_0 L/\lambda)[1 - (1/2)\gamma\, n_0^2 E_z] = \phi_0 - \Delta\phi \tag{8.34}$$

The net phase shift or total phase retardation between these two components can be written as

$$\phi = \phi_x - \phi_y = (2\pi L/\lambda)(\gamma\, n_0^3 E_z) = (2\pi/\lambda)(\gamma\, n_0^3 V\,) \tag{8.35}$$

The electric field components of the optical wave emerging from the electro-optic crystal are

$$E_x = E_0/\sqrt{2}[\cos(\omega t + \phi_0 + \Delta\phi)]$$

$$E_y = E_0/\sqrt{2}[\cos(\omega t + \phi_0 - \Delta\phi)] \tag{8.36}$$

The analyzer is nothing but a polarizer whose optical axis is oriented horizontally. The resultant electric field, when the optical beam emerges from the analyzer, is thus given by

$$E_H = |E_x/\sqrt{2}| + |E_y/\sqrt{2}| \tag{8.37}$$

The resultant light beam intensity can be obtained by integrating the square of the electric field over the complete period. Thus

$$\begin{aligned} I &= (\omega/2\pi)\int E^2 dt \\ &= I_0 \sin^2(\Delta\phi) = I_0 \sin^2(\phi/2) \end{aligned} \tag{8.38}$$

where I_0 is the incident intensity and ϕ is the total phase retardation between the polarized waves. Since ϕ is a function of the applied voltage V [see Eq. (8.35)], the output intensity can be controlled by this voltage. Thus I/I_0 can be plotted as a function of voltage as shown in Fig. 8.9.

The transmission characteristic is varied in determining the bias voltage that has to be applied to the electro-optic modulator before the signal is injected into it. It can be seen from this characteristic that at $V = 0$, the optical input is completely blocked by the modulator. This is easy to explain because, when the polarizer vertically polarizes the input light, it will be completely blocked by the analyzer which is a horizontal polarizer. When the applied voltage V across the electro-optic crystal is increased, the net phase of the carrier increases and the horizontal component of the intensity which passes through the analyzer also increases. The transmitted intensity of the optical carrier will reach a maximum when the net phase introduced between the ordinary and extraordinary components reaches π radians. This would mean that the resultant phase of the optical carrier would be rotated by

$\pi/2$ radians. Hence, the optical carrier will be allowed to pass unaffected by the analyzer. The transmission characteristic can be described by the following equation

$$I/I_0 = \sin^2[(\pi/2)(V/V_0)] \tag{8.39}$$

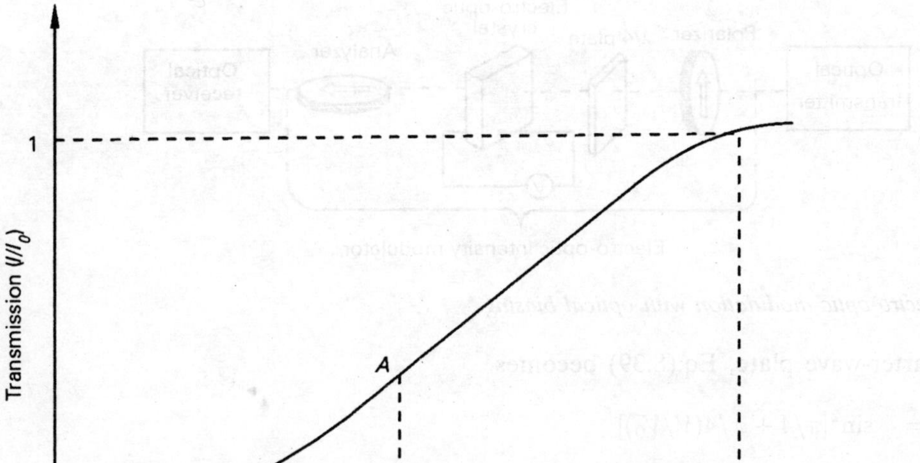

Fig. 8.9 *Transmission characteristics of electro-optic modulator.*

Equation (8.39) will explain the behaviour of the electro-optic modulator. The transmission characteristic is nonlinear at all points except at *A*. Thus, in order to operate the device with minimum distortion, a d.c. bias voltage of $V_0/2$ should be applied. The value of V_0 [in Eq.(8.39)] is a function of the electro-optic material used as the modulator, and can be deduced from Eq.(8.35) by putting $\phi = \pi$ which results in

$$V_0 = (\lambda/2\gamma n_0^3) \tag{8.40}$$

Example 8.1 A KDP crystal is used in an electro-optic modulator in order to modulate a He–Ne laser beam externally. Calculate: (i) the half-wave retardation voltage V_0, if the refractive index of KDP is 1, and (ii) Pockels electro-optic coefficient of this crystal.

Solution For a He–Ne laser, $\lambda = 632.8\ \mu$m, $n_0 = 1.512$, and $\gamma = 10.5 \times 10^{-12}$ m/V. Hence

$$V_0 = (\lambda/2\gamma n_0^3)$$

$$V_0 = \frac{632.8 \times 10^{-6}}{2 \times 10.5 \times 10^{-12} \times (1.512)^3} = 8717.469 \text{ kV}$$

From the above example, it can be seen that the bias voltage for KDP and its class of crystals runs into kilovolts, making it cumbersome and expensive to drive it with a signal. One way of overcoming it is by biasing it optically by introducing a quarter-wave ($\lambda/4$) plate in it as shown in Fig. 8.10.

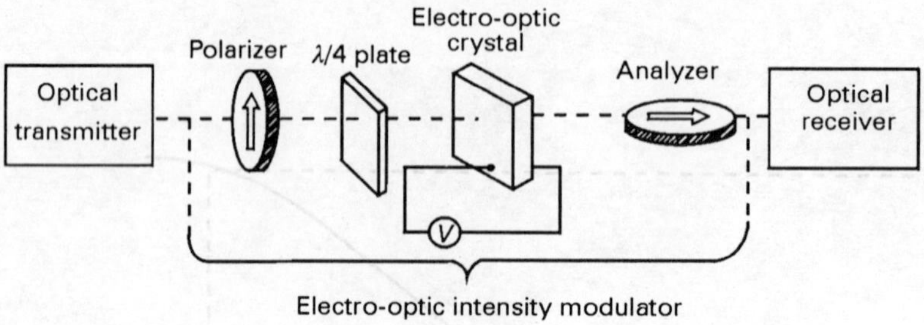

Fig. 8.10 *Electro-optic modulation with optical biasing.*

With the quarter-wave plate, Eq.(8.39) becomes

$$\begin{aligned} I/I_0 &= \sin^2[\pi/4 + \pi/4(V/V_0)] \\ &= \frac{1}{2}\left\{1 + \sin\left[\left(\frac{\pi}{2}\right)\left(\frac{V}{V_0}\right)\right]\right\} \end{aligned} \tag{8.41}$$

For small signals, when $V/V_0 \ll 1$, Eq.(8.41) reduces to

$$I/I_0 = 0.5 + (\pi/4)(V/V_0) \tag{8.42}$$

If the modulating signal frequency is f_m and the depth of modulation is $m = V/V_0$, then

$$V/V_0 = \sin(\omega_m t) \quad \text{and} \quad I/I_0 = 0.5 + (\pi m/4)\sin(\omega_m t) \tag{8.43}$$

and the amplitude modulation of the emitted radiation occurs. However, since $m \ll 1$, the modulated or a.c. component of the optical signal is much smaller than its d.c. component. In order to extend the range of operation of the electro-optic modulator into microwave frequencies, it is possible to design it in the form of a microwave resonator cavity, in which the optical components are implanted. Lithium-based electro-optic materials like lithium niobate and lithium tantalate are very attractive for modulation since their optical properties are much better, and the required half-wave retardation voltage is much smaller, which is of the order of a few hundreds of volts only.

In the transverse mode of operation, the external signal voltage V is applied to the electro-optic crystal in such a way that the resulting electric field is perpendicular to the direction of propagation of the optical carrier through it. The half-wave retardation voltage V_0 in this case can be deduced on the same lines as mentioned above, and is given by

$$V_0 = \frac{\lambda_0}{n_0^3\gamma}\left[\frac{d}{L}\right] \tag{8.44}$$

where d is the breadth of the crystal across which the external voltage is applied. From Eq.(8.44), it is obvious that V_0 can be reduced by decreasing the (d/L) ratio. However, there are limitations to decreasing the (d/L) ratio. The length L of the modulator crystal is limited to a few centimetres due to problems associated with growing large crystals. The depth d of the crystal is limited by a Gaussian beam having a minimum waist radius w_0, which passes through the crystal of length l (Fig. 8.11). Within these limitations, the lowest possible half-wave retardation voltage for the transverse mode in the case of KDP crystal is 3.4 kV compared to 8.3 kV for the longitudinal mode. In the case of lithium-based crystals, these voltages will be correspondingly smaller.

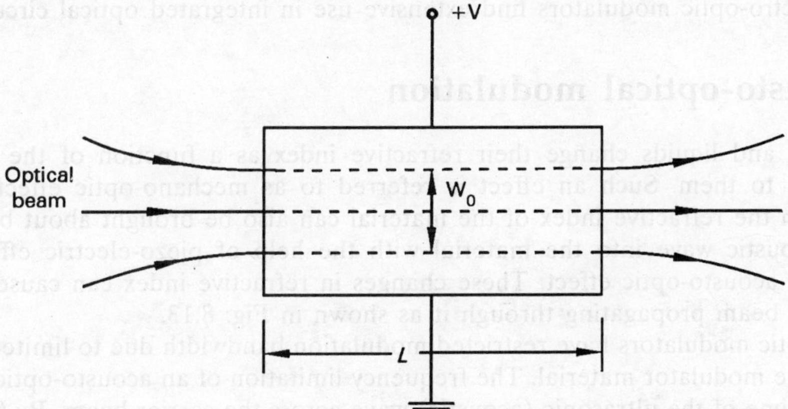

Fig. 8.11 The geometry of the optical beam passing through the electro-optic crystal.

Thus, it is possible to design a modulator or switches using low voltages. A strip waveguide is formed in an electro-optic material like lithium tantalate by the diffusion of titanium. When a voltage is applied across it, it will change the effective refractive index,

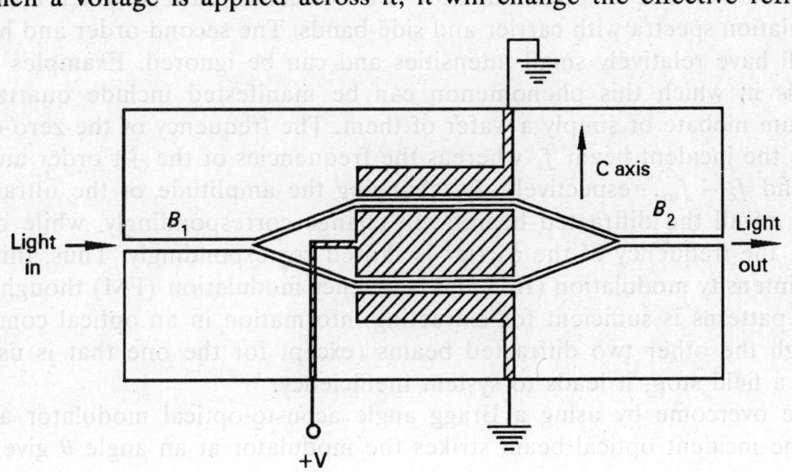

Fig. 8.12 An integrated optic amplitude modulator using the Mach–Zehnder interferometer configuration.

thus changing the phase velocity of the optical beam propagating through it. An integrated optic amplitude modulator using the Mach–Zehnder interferometer configuration is shown in Fig. 8.12. A single-mode waveguide splits into two single-mode waveguides which, after a certain length, rejoin to form a single-mode waveguide. Oppositely-directed electric fields are applied on the two arms of the interferometer by the electrode pattern as shown in the figure. Thus, a voltage V applied at the electrodes creates a phase shift of $\Delta\phi/2$ and $-\Delta\phi/2$ on the light propagating through the two arms, leading to a total phase-shift of $\Delta\phi$. The beam of light from each arm interferes at the output, leading to amplitude modulation in proportion to the applied voltage. One half-cycle of intensity modulation can be typically achieved in such a case with a voltage as low as 1.6 V. These types of miniature electro-optic modulators find extensive use in integrated optical circuits.

8.6 Acousto-optical modulation

Some crystals and liquids change their refractive index as a function of the mechanical strain applied to them. Such an effect is referred to as mechano-optic effect. However, the changes in the refractive index of the material can also be brought about by launching a pressure/acoustic wave into the material with the help of piezo-electric effect. This is referred to as acousto-optic effect. These changes in refractive index can cause diffraction of a light was beam propagating through it as shown in Fig. 8.13.

Acousto-optic modulators have restricted modulation bandwidth due to limited frequency response of the modulator material. The frequency limitation of an acousto-optic modulator is the transit time of the ultrasonic (acoustic) wave across the carrier beam. By focusing the laser beam to a narrow diameter, a fairly large modulation bandwidth is achievable. The ultrasonic wave propagating in the vertical direction creates a grating in the modulating material which is transparent to the optical beam propagating through it in the horizontal direction. This erasable/changeable grating whose line-spacing corresponds to the ultrasonic wavelength λ_s, diffracts the optical beam propagating through it as shown in Fig. 8.14. The resultant on the other side of the modulator is a diffraction pattern which corresponds to the AM modulation spectra with carrier and side-bands. The second-order and higher order side-bands will have relatively small intensities and can be ignored. Examples of acousto-optic materials in which this phenomenon can be manifested include quartz, cadmium sulphide, lithium niobate or simply a wafer of them. The frequency of the zero-order beam is the same as the incident beam f_c whereas the frequencies of the +1 order and −1 order are $f_c + f_m$ and $f_c - f_m$, respectively. By changing the amplitude of the ultrasonic wave, the intensities of all the diffracted beams will change correspondingly, while by changing the frequency, the frequency of the carrier is shifted correspondingly. Thus, this technique may result in intensity modulation (IM) and frequency modulation (FM) though any one of the diffracted patterns is sufficient for extracting information in an optical communication system. Though the other two diffracted beams (except for the one that is used) can be eliminated by a field stop, it leads to system inefficiency.

This can be overcome by using a Bragg angle acousto-optical modulator as shown in Fig. 8.14. If the incident optical beam strikes the modulator at an angle θ given by

$$\theta = \sin^{-1}(\lambda_c/2\lambda_s) \tag{8.45}$$

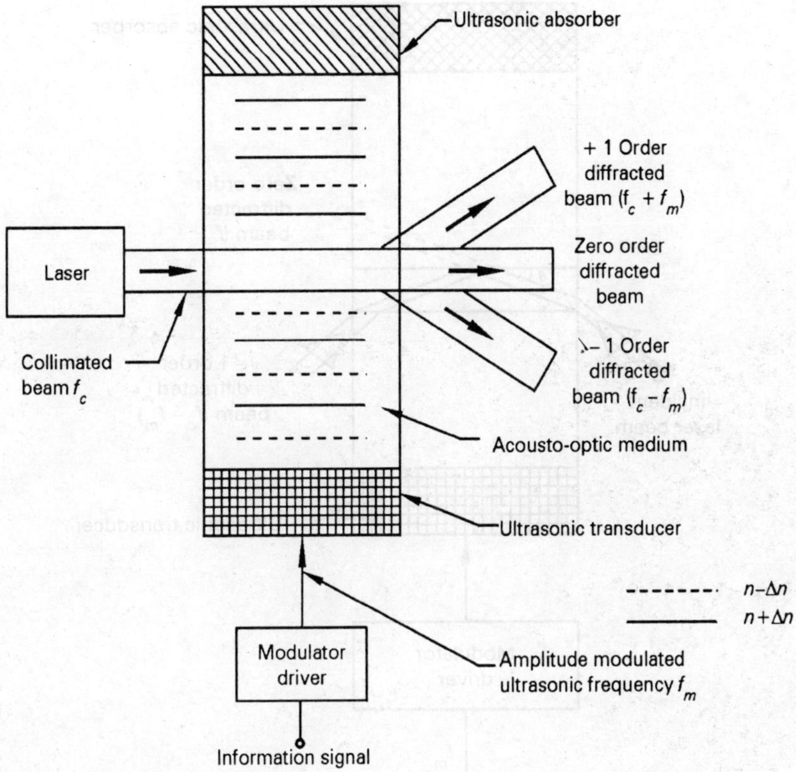

Fig. 8.13 *Basic acousto-optic modulator.*

(θ is commonly known as the Bragg angle), then the +1-order diffracted beam is eliminated. The intensities of the 0-order and −1-order diffracted beams, which will be correspondingly higher, are given by

$$I_0 = \cos^2(R_N/2) \quad \text{and} \quad I_{-1} = \sin^2(R_N/2) \tag{8.46}$$

where R_N is referred to as Raman–Nath parameter and is given by the equation

$$R_N = \frac{2\pi\, w_u \Delta n}{\lambda_c} \tag{8.47}$$

Here w_u is the width of the ultrasonic cell and Δn is the change in the index of refraction of the medium caused by the ultrasonic wave. For sufficiently large values of R_N, approximately 95% of the intensity of the incident laser beam may be transferred to the −1-order diffracted beam. Since the Raman–Nath parameter is inversely proportional to λ_c, acousto-optical modulators are not as effective at longer wavelengths. With Bragg angle acousto-optical modulation, by transferring most of the carrier wave energy to the first-order diffracted beam, FM modulation can be very efficient. Besides, it offers a means of laser frequency shifting which is useful for tuning a local oscillator for heterodyne detection.

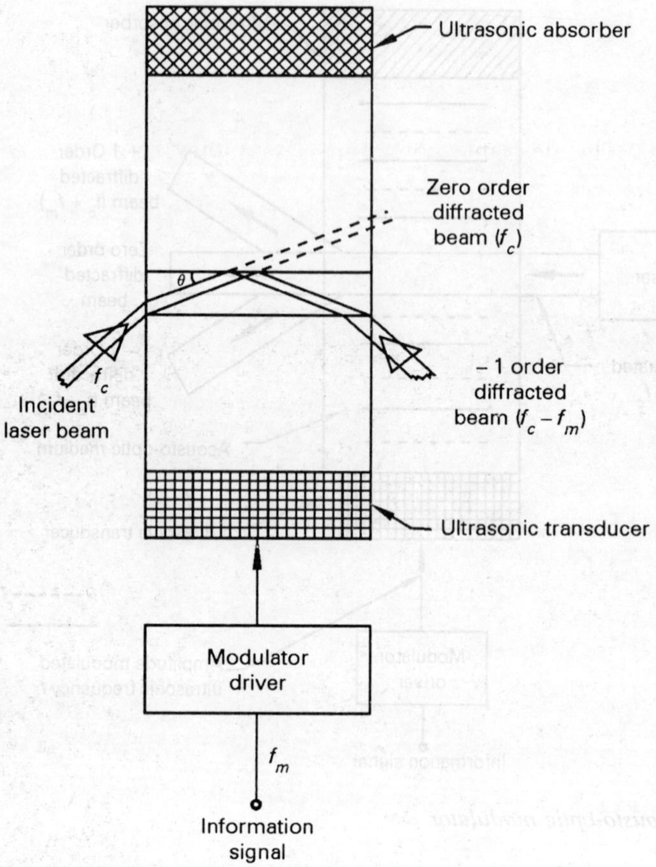

Fig. 8.14 Bragg angle acousto-optic modulator/deflector.

8.7 Modulation of semiconductor devices

Most fibre-optic communication systems employ semiconductor devices like laser diodes (LDs) and light emitting diodes (LEDs) as transmitters. In many cases, where short-haul communication links are involved, it is possible to bridge the distances with the lower optical power of the LEDs. LEDs offer many advantages over LDs in terms of simplicity of driving circuits, cost and relative ease with which LEDs can be operated. Of course, they lack the power to carry information over long distances. Also, because of lack of spectral purity, they cannot be frequency or phase modulated. However, since from the receiver point of view intensity modulation (IM) is preferred both for LEDs and LDs, we will consider the possibilities of driving currents through these devices as a function of information signal.

8.7.1 Modulation circuits for light emitting diodes

It is possible to use different circuits for energizing LEDs in order to emit optical carrier/radiation from them either in an analog mode or a digital mode as a function of the information signal. The desirable condition is that they should be as fast as possible in order to handle the high-frequency/bit rate information signal. The ultimate limit is the speed of LEDs themselves. The speed of LEDs is conditioned by the mobility of the charge carriers that produce photons. By dramatic improvements in device technologies, the optical communication grade LEDs can now handle GHz/Gbit rate signals. Thus, the electronic drive circuit designers have now developed suitable circuits which can match the speed of LEDs. In response to this, an area now referred to as high speed electronics, based on fast electronic devices, has emerged.

The optical power generated by an LED is linearly proportional to the forward driving current. Hence, digital modulation of LEDs can be done by injecting a series of current pulses through the device as shown in Fig. 8.15. This can be achieved, for example, by using

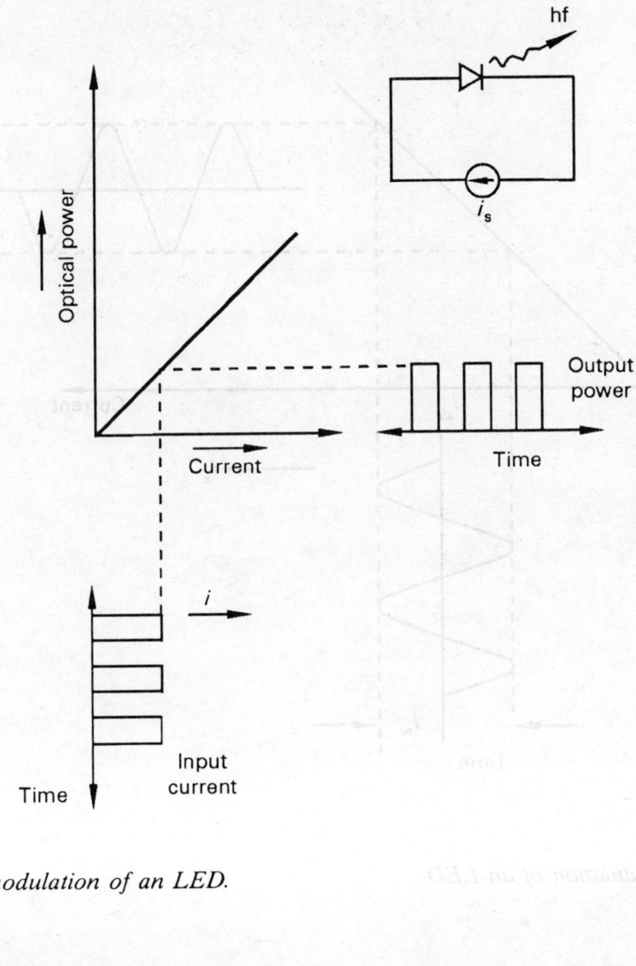

Fig. 8.15 Digital modulation of an LED.

an electronic circuit as shown in Fig. 8.15. For analog modulation, a bias current I_{DC} before the signal current i_s is supplied to the LED as shown in Fig. 8.16. This can be achieved easily by modifying the earlier transistor digital driver circuit by suitably providing the bias voltage to the base. We have already seen that the light output from LEDs has a large divergence angle which makes it rather difficult to launch the energy into the fibre. A fairly large portion of the emitted optical radiation may not be coupled into the fibre by usual means, especially when low NA fibres are involved. In this aspect, LDs are far better.

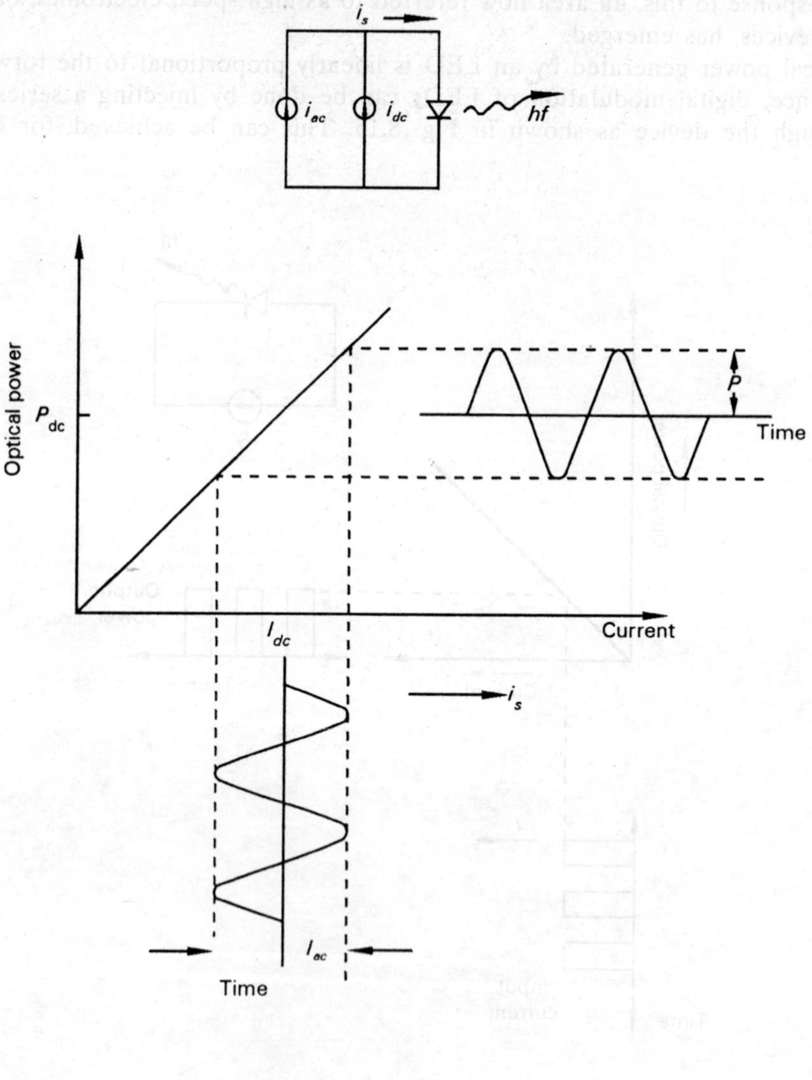

Fig. 8.16 Analog modulation of an LED.

8.7.2 Laser diode modulation circuits

Laser diodes present more problems to the circuit designer than LEDs. This is due to the fundamental difference in the fabrication/operation between these two devices. The LD has the following characteristics which have to be taken into account while designing the driver circuit:

(i) The existence of a threshold current
(ii) The threshold current's age dependence
(iii) The threshold current's temperature dependence
(iv) The emission wavelength's temperature dependence

In order to ensure a constant power/frequency output, a cooler is provided inside the laser diode housing as shown in Fig. 8.17. This ensures a stable output. The cooler is driven by a built-in photodetector-based circuit so that the bias control is internally adjusted as shown in the figure.

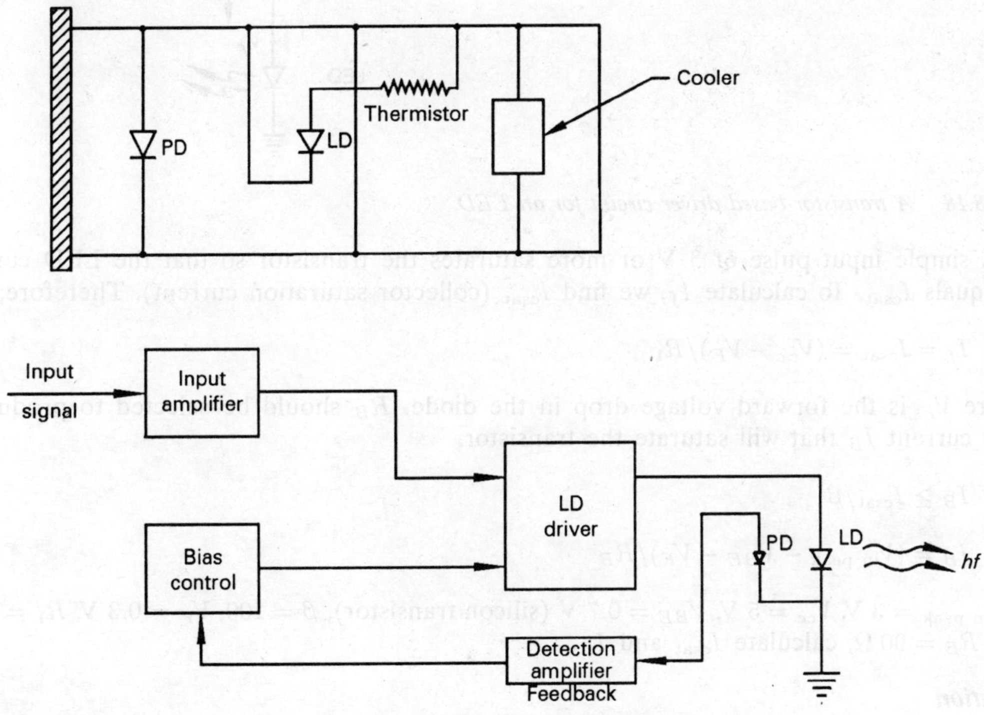

Fig. 8.17 The integrated laser diode package which includes a cooler.

Because of the existence of a threshold current, even for digital signals, the device has to be suitably biased. Since the overall electrical power requirements for energizing an LD are higher than LEDs, suitable circuit elements have to be employed. LDs offer much higher bandwidth and larger power outputs than LEDs. Their coupling efficiency with the fibre is also far better. Hence they are usually employed in long-haul optical communication links.

Example 8.2 *Practical modulation circuit* In practice, transistors are often used as the switching elements for LEDs and LDs, provided the switching speed of the transistor is compatible with the rise-time requirements. In general, the LED switching circuits are simpler because smaller currents are involved. A typical transistor-based driver circuit for an LED and its operation is shown below (see Fig. 8.18).

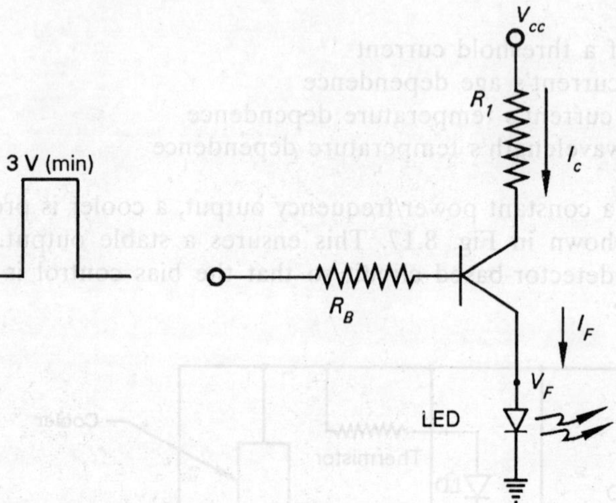

Fig. 8.18 A transistor-based driver circuit for an LED.

A simple input pulse of 3 V or more saturates the transistor so that the LED current I_f equals I_{csat}. To calculate I_f, we find I_{csat} (collector saturation current). Therefore,

$$I_f = I_{\text{csat}} = (V_{cc} - V_F)/R_1.$$

where V_F is the forward voltage drop in the diode. R_B should be selected to produce a base current I_B that will saturate the transistor.

$$I_B \geq I_{\text{csat}}/\beta$$

$$I_B = (V_{\text{in peak}} - V_{BE} - V_F)/R_B$$

If $V_{\text{in peak}} = 3$ V, $V_{cc} = 5$ V, $V_{BE} = 0.7$ V (silicon transistor), $\beta = 100$, $V_F = 0.3$ V, $R_1 = 45\ \Omega$ and $R_B = 90\ \Omega$, calculate I_{csat} and I_B.

Solution

$$I_B = \frac{3\text{V} - 0.7 - 0.3}{90} = 2/90 = 22 \text{ mA}$$

$$I_{\text{csat}} = \frac{(V_{cc} - V_F)}{R_1} = \frac{5 - 0.3}{45} = 104 \text{ mA}.$$

Therefore, $I_B \geq I_{\text{csat}}/\beta$ for the device (LED) to emit pulse of optical energy.

9 Optical communication systems

9.1 Historical background

Because of the well-known and obvious difficulties encountered by the early open-air atmospheric optical communication links, fiber-optic communication links have emerged in the 1970s. These have been steadily improved upon, in terms of both range and information-carrying capacity. Conventional communication links around the world are either replaced or augmented by optical communication systems at a hectic pace. With the introduction of trans-oceanic fiber-optic cables, a debate has been raging whether the expensive microwave satellite global communication has become redundant. In spite of the many advances in the fiber-optic communication systems, open-air atmospheric optical links are still in use in special situations. Besides, satellite optical/laser communication systems are under development and are likely to be deployed soon.

The first generation fiber-optic communication links used sources at 850 nm which operated with GaAs sources. Though this did not correspond to the minimum in the fiber attenuation, they still could exploit the peak response of the silicon-based photodetectors around this wavelength (see Fig. 9.1). The major components of a fiber-optic

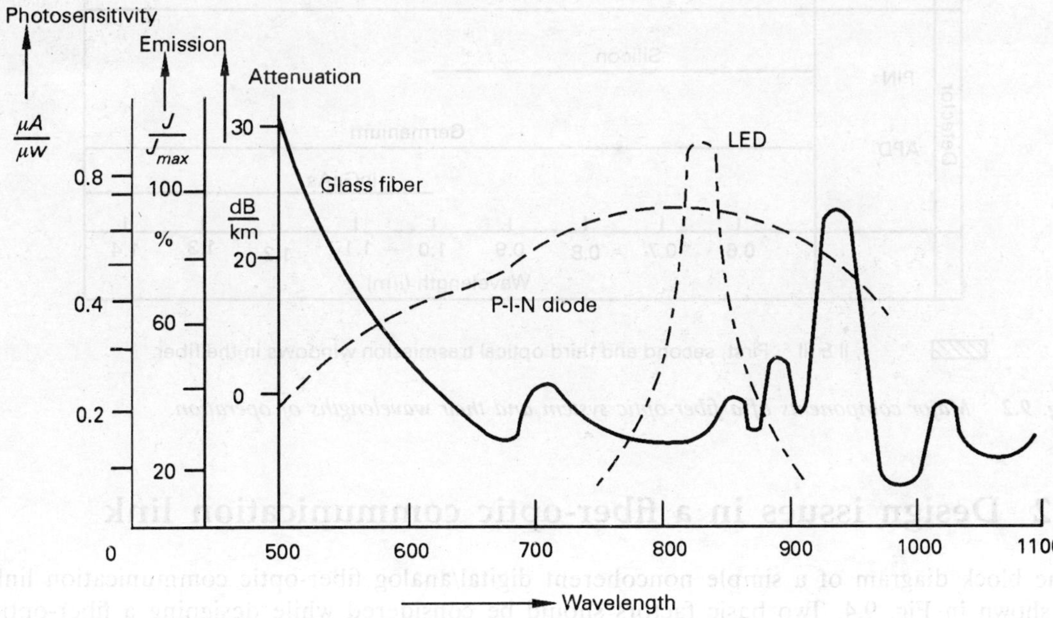

Fig. 9.1 Attenuation vs wavelength characteristics for glass fiber, P-I-N diode and LED.

system and their wavelengths of operation are shown in Fig. 9.2. Once the advantages of shifting the wavelength of operation to 1300 nm were discovered, both sources and detectors were designed and fabricated using InGaAsP materials, which gave peak performance at these wavelengths. This, in conjunction with a graded-index optical fiber started yielding high-bit rates. However, the real push towards large-information handling was made with the advent of the wavelength-division multiplexing (WDM) process which was invented and patented by Norman French in 1934. The basic ideas of WDM are given in Fig. 9.3.

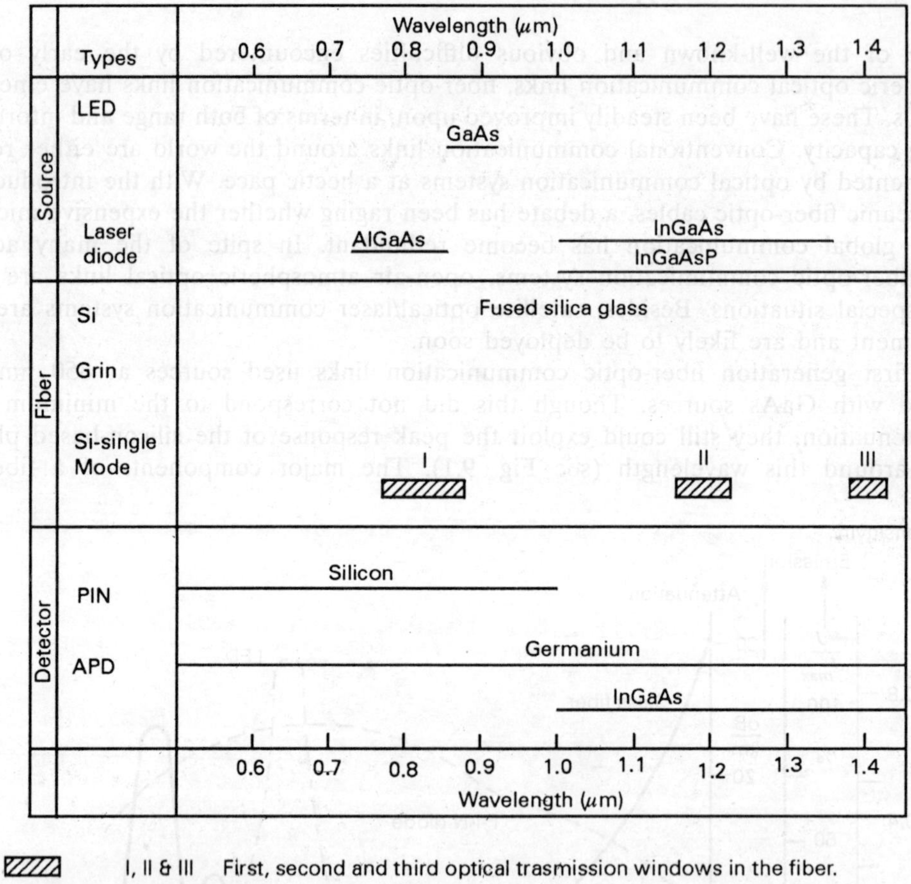

Fig. 9.2 Major components of a fiber-optic system and their wavelengths of operation.

9.2 Design issues in a fiber-optic communication link

The block diagram of a simple noncoherent digital/analog fiber-optic communication link is shown in Fig. 9.4. Two basic factors should be considered while designing a fiber-optic communication link. One of them is the optical power budgeting in the link, which will help in determining the optimum repeaterless optical communication length or range. The second factor is the time budgeting which gives us an insight into the usable bandwidth of such a link.

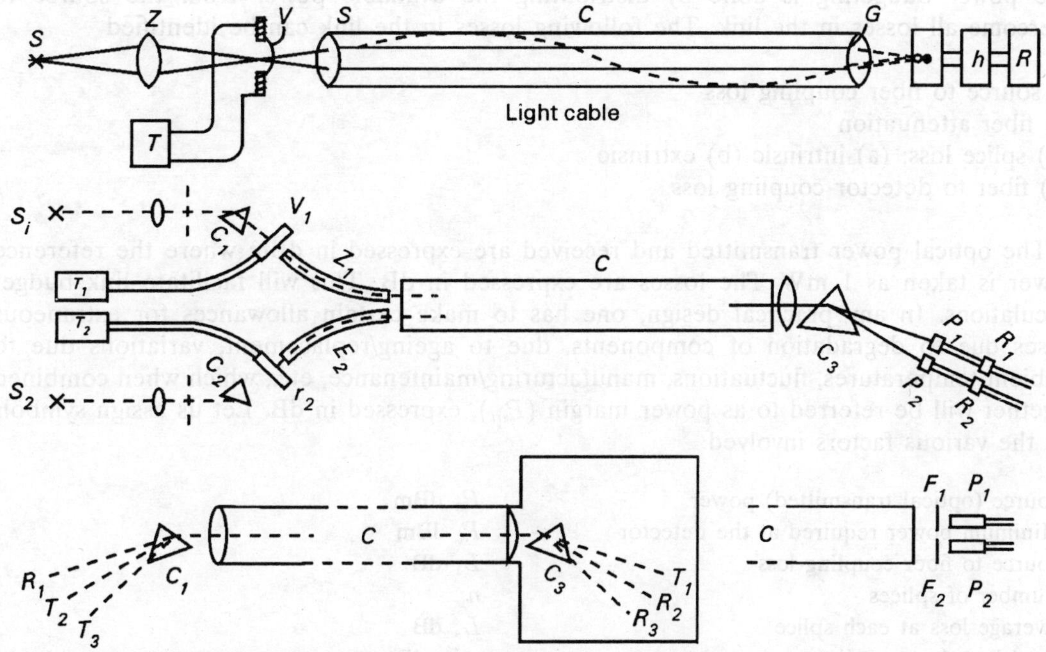

Fig. 9.3 Basic idea of WDM according to Norman French in 1934.

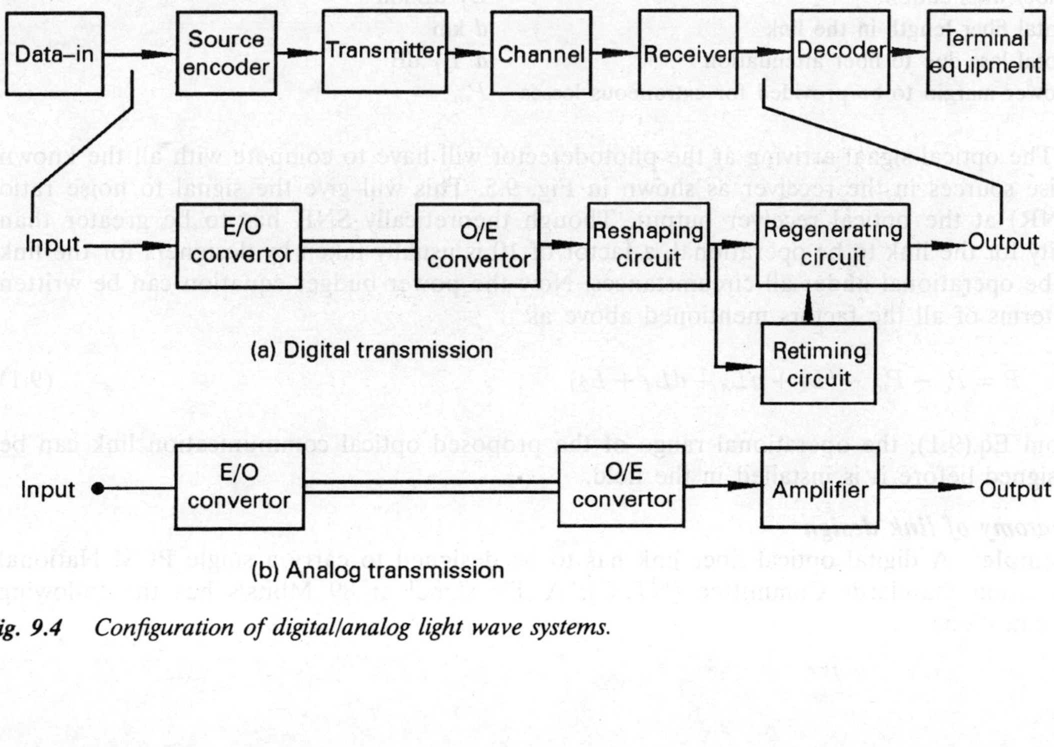

Fig. 9.4 Configuration of digital/analog light wave systems.

9.2.1 Power budget

The power budgeting is done by distributing the available power from the source to overcome all losses in the link. The following losses in the link can be identified

(i) source to fiber coupling loss
(ii) fiber attenuation
(iii) splice loss: (a) intrinsic (b) extrinsic
(iv) fiber to detector coupling loss.

The optical power transmitted and received are expressed in dBm where the reference power is taken as 1 mW. The losses are expressed in dB. This will facilitate link budget calculations. In any practical design, one has to make certain allowances for extraneous losses due to degradation of components, due to ageing/replacement, variations due to ambient temperatures, fluctuations, manufacturing/maintenance, etc., which when combined together will be referred to as power margin (P_m), expressed in dB. Let us assign symbols for the various factors involved.

Source (optical transmitted) power	P_o dBm
Minimum power required at the detector	P_i dBm
Source to fiber coupling loss	L_1 dB
Number of splices	n
Average loss at each splice	L_s dB
Total loss due to splices	nL_s dB
Fiber to detector coupling loss	L_2 dB
Fiber attenuation	L_f dB/km
Total fiber length in the link	d km
Total loss due to fiber attenuation	$d^* L_f$ dB
Power margin to be provided for extraneous losses	P_m

The optical signal arriving at the photodetector will have to compete with all the known noise sources in the receiver as shown in Fig. 9.5. This will give the signal to noise ratio (SNR) at the optical receiver output. Though theoretically SNR has to be greater than unity for the link to be operational, a factor of 10 is usually taken by designers for the link to be operational under all circumstances. Now the power budget equation can be written in terms of all the factors mentioned above as

$$P = P_o - P_m - (L_1 + nL_s + dL_f + L_2) \tag{9.1}$$

From Eq.(9.1), the operational range of the proposed optical communication link can be designed before it is installed in the field.

Anatomy of link design

Example: A digital optical fiber link has to be designed to carry a single PCM National Television Standards Committee (NTSC). A TV signal at 89 Mbits/s has the following specifications

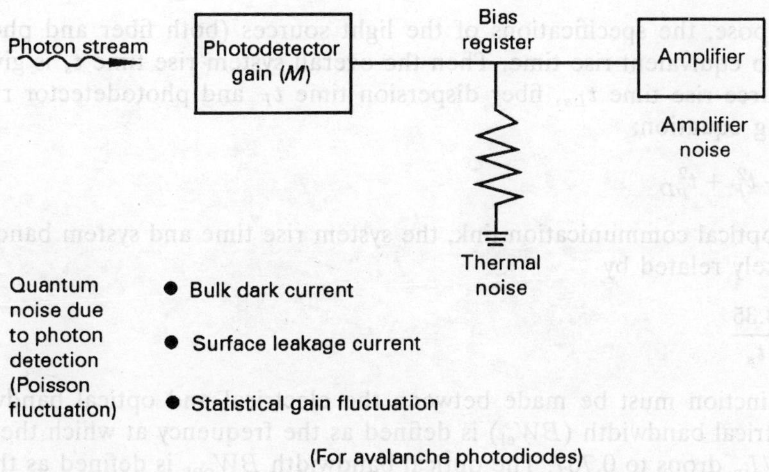

Fig. 9.5 *Noise sources and disturbances in the optical pulse detection mechanism.*

Link span length	10 km at 100 Mbit/s.
Laser transmitter output into 50 μm core graded-index fiber at 0.85 μm	0 dBm.
Receiver sensitivity	50 dBm at 100 Mbit/s.
Fiber loss	3 dB/km.
Fiber dispersion	10 ns over 10 km.

Link budget

Transmitter output	0 dBm
Receiver sensitivity	−50 dBm for SNR of 20 dB
Margin	50 dBm
Fiber loss, 3 dB/km	30 dB
Total splice loss	3 dB
Total connector loss	4 dB
Detector coupling loss	1 dB
Head room for temperature change and ageing and future splices	9 dB
Total attenuation	47 dB
Excess power margin	3 dB
Fiber dispersion	10 ns over 10 km.

9.3 Time budget

The next important issue in the design of a fiber-optic communication link is the time budget which leads to the computation of the overall bandwidth capability of the system.

For this purpose, the specifications of the light sources (both fiber and photodetector) are converted to equivalent rise time. Then the overall system rise time t_s is given in terms of the light source rise time t_{Ls}, fiber dispersion time t_F and photodetector rise time t_{pD} by the following equation:

$$t_s^2 = t_{Ls}^2 + t_F^2 + t_{pD}^2 \tag{9.2}$$

In a digital optical communication link, the system rise time and system bandwidth BWd are approximately related by

$$BWd = \frac{0.35}{t_s} \tag{9.3}$$

Here, a distinction must be made between the electrical and optical bandwidth of the fiber. The electrical bandwidth (BW_{el}) is defined as the frequency at which the ratio of the intensities I_{out}/I_{in} drops to 0.707. The optical bandwidth BW_{opt} is defined as the frequency at which the ratio P_{Lo}/P_{Li} drops to 0.5. This leads to the relation

$$BW_{el} = 0.707 BW_{opt} \tag{9.4}$$

The equivalent fiber dispersion time is then given by

$$t_F = \frac{0.35}{BW_{el}} \tag{9.5}$$

The fiber dispersion time can be estimated from the above two equations. First, the fiber optical bandwidth is estimated from the fiber dispersion measurements. As we have already seen earlier, the pulse dispersion in a fiber is predominantly due to modal dispersion and material dispersion which can be estimated. The rise time of the light source is contributed by two factors:

(i) the mobility of the charge carriers (t_{cc}) in it which will determine the speed with which the photons are generated and
(ii) the rise time due to the driver circuit (t_{dc}) which activates the light source.

Finally, the rise time of the photodetector is due to the transit time effects (t_{TR}) and the contribution due to electronic receiver circuit (t_{Rc}) used along with it. The following example will give a feel for the numbers involved.

Example 9.1 In a 2 km optical fiber communication link, the optical bandwidth per kilometre of the fiber is given as 25 MHz. The rise time due to the mobility of the charge carrier is estimated to be 10 ns and the rise time of the driver circuit is given as 2 ns. On the receiver side, the rise time due to the transmit time effects in the photodetector is estimated to be 3 ns and the contribution due to electronic receiver circuits employed is 12 ns. Calculate an overall estimate of the system bandwidth.

Solution Optical bandwidth of the 2 km fiber $= BW_{opt} = \frac{25}{2} = 12.5$ MHz

Electrical bandwidth of the fiber $= 0.707 BW_{opt} = 8.8375$ MHz

Thus the fiber dispersion time $= t_F\left(\dfrac{0.35}{8.8375\times10^6}\right) = 39.6$ ns

The overall system rise time $= t_s = \sqrt{(t_{cc}^2 + t_{dc}^2 + t_f^2 + t_{TR}^2 + t_{RC}^2)}$

$$= \sqrt{10^2 + 2^2 + (39.6)^2 + 3^2 + 12^2} = 42.7 \text{ ns.}$$

The electrical system bandwidth can now be calculated as

$$BW_{\text{el}} = \frac{0.35}{42.7\times10^{-9}} = 8.2 \text{ MHz.}$$

In the above example, note that the fiber itself has the highest dispersion namely 39.6 ns. Thus the total bandwidth of the system is almost entirely due to the fiber. Improving the fiber bandwidth (using low dispersion fibers) is, therefore, the key to improving the overall system bandwidth. Thus, in analyzing and designing large bandwidth optical communication links, the worst component in terms of its rise time should always be improved first.

9.4 Optical repeater and amplifiers

We have seen so far that the span length and its information carrying capability are limited by losses in the fiber and its dispersion respectively. However, in long-distance optical communication links, it is necessary to periodically restore the stream of optical pulses to their original magnitude and shape. Thus, the maximum repeater loss length of a fiber is determined by the fiber attenuation and its dispersion. Naturally, one would obviously like to reduce the necessity for the number of repeaters in a long-haul optical communication link as it would reduce both the cost and the complexity of the overall system. Further, each repeater demands electrical power input in order to do its job and this might sometimes prove to be difficult—especially when the link is passing through an inaccessible terrains like trans-oceanic links. The block diagram of an electric repeater in a fiber-optic communication link is shown in Fig. 9.6(a). The electronics in the box as shown in the figure undertakes the job of reshaping, retiming and regeneration of the pulse stream. At high-bit rates, the electronics employed must be high-speed calling for fundamental changes in the electronic circuitry and devices. Recent developments in high-speed electronics have met this challenge magnificently.

Technological breakthroughs which lead to the development of near-zero dispersive fibers, and the fabrication of narrow line-width semiconductor lasers have resulted in eliminating the necessity for reshaping and retiming of the optical pulses even at high-bit rates. Thus, the task of the electronics in the repeater is now reduced (simplified) to amplifying the pulse stream at periodic intervals. However, the necessity to deploy a photodetector and a laser in the repeater still remains. This can only be eliminated if an optical amplifier is deployed in place of an electronic repeater. Such an optical amplifier has been developed in recent years by doping a stretch of the fiber with erbium (Er). When optically pumped, this stretch will act like an optical amplifier due to the population inversion created in it. A typical block diagram of an erbium-doped fiber amplifier (EDFA) is shown in Fig. 9.6(b). Sometimes, semiconductor laser amplifiers (SCLAs) or Raman optical fiber amplifiers (RFAs) are used for the same purpose. However, EDFAs are simplest and easiest to install, and because of their simplicity they are relatively inexpensive and reliable. Further, they can be easily adopted for multichannel operation and can handle very high data rates. Thus, in recent long-haul optical communication links, optical amplifiers are increasingly being deployed in order to reduce the number of repeaters.

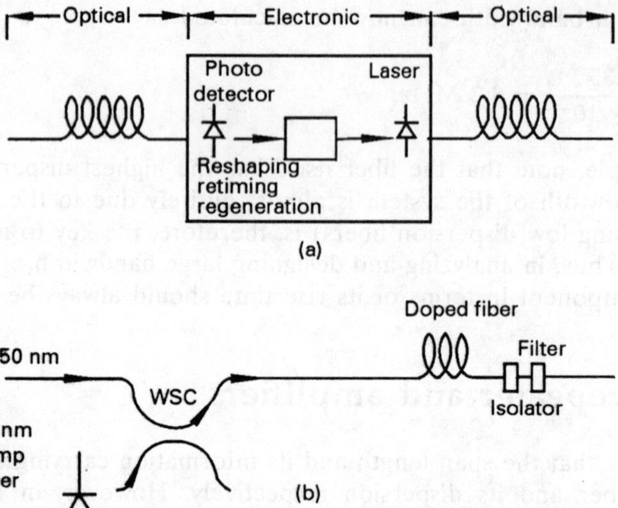

Fig. 9.6 (a) Block diagram of an electronic repeater in a fiber-optic communication link (b) Block diagram of an erbium-doped fiber amplifier (EDFA).

9.5 Coherent optical systems

The emergence of coherent optical communication is similar to the development of superheterodyne techniques in radio communication. The advantages offered by coherent optical communication are also analogous to those in radio communication namely, better sensitivity and fine tunability. This, in other words, means better utilization of the inherently large bandwidth available in a single-mode optical fiber communication. Research in this field for more than a decade has now resulted in impressive developments in recent years. Over the past several years, optical coherent communication has evolved from an esoteric subject studied in a few research laboratories and has moved into the field for deployment. This has provided real gains in terms of receiver sensitivity, resulting in an increased repeaterless transmission span over what was achieved using the intensity modulated direct detection systems. Moreover, the great selectivity of coherent receivers can also be utilized for subscriber network applications, where at the user terminal stage, it might become possible to tune the receiver to the designed channel.

The most general experimental arrangement for a coherent heterodyne optical communication system is shown in Fig. 9.7. The optical transmitter in the coherent system consists of a narrow line-width single-mode semiconductor laser diode which is typically a distributed feedback (DFB) laser diode having a line-width of approximately 10–100 MHz. An external cavity/grating laser diode having a spectral line-width of 10–100 kHz is used where higher sensitivity in the communication system is desired. A multimode laser or LED cannot be used in this system because the source line-width must be narrower than the signal bandwidth. Since the frequency stability of the source is of prime importance in coherent systems, the laser source is often placed in a temperature-stabilized chamber whose variation is controlled within ± 0.01°C.

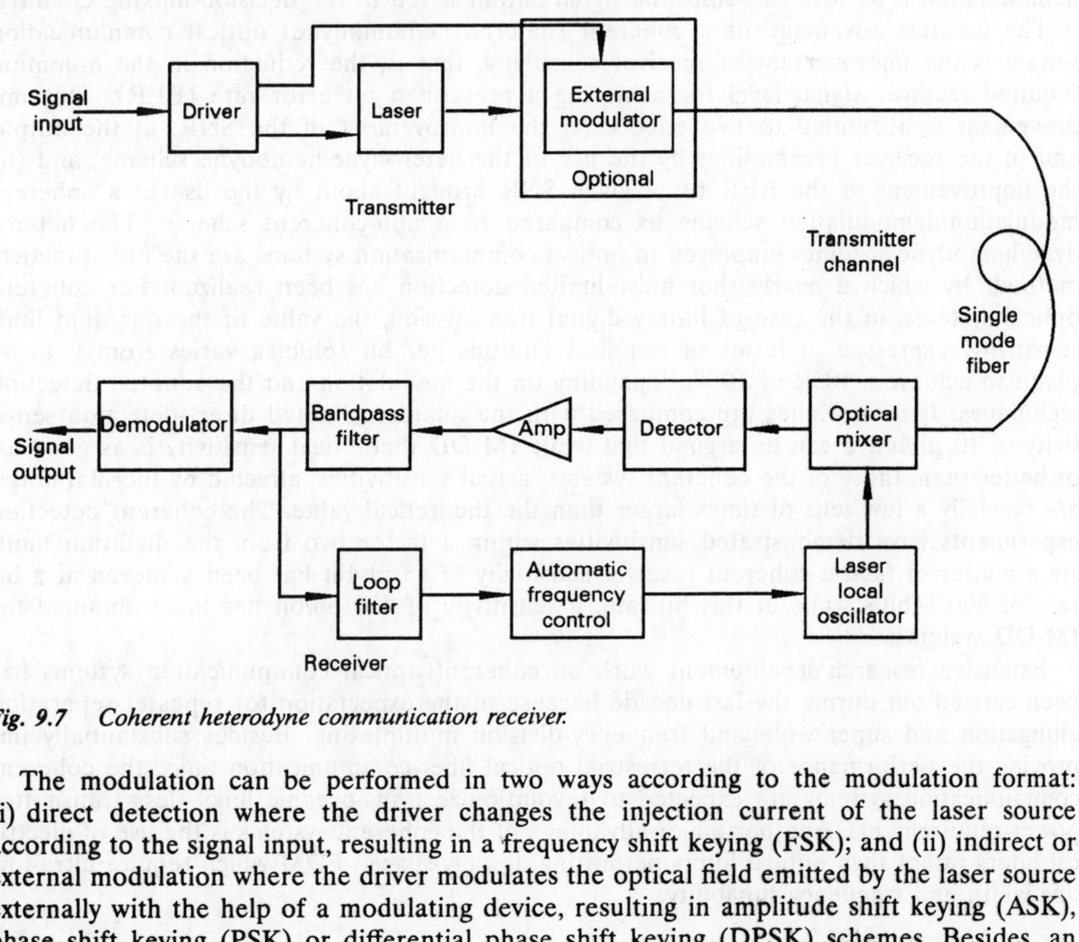

Fig. 9.7 Coherent heterodyne communication receiver.

The modulation can be performed in two ways according to the modulation format: (i) direct detection where the driver changes the injection current of the laser source according to the signal input, resulting in a frequency shift keying (FSK); and (ii) indirect or external modulation where the driver modulates the optical field emitted by the laser source externally with the help of a modulating device, resulting in amplitude shift keying (ASK), phase shift keying (PSK) or differential phase shift keying (DPSK) schemes. Besides, an optical isolator may have to be placed between the source and the fiber in order to avoid back reflections, which could cause frequency instability of the laser source. An external modulator is particularly useful when phase or polarization modulation is desired. However, the use of an external modulator introduces additional insertion losses in the link. The direct modulation technique is often used because it is less complex and relatively inexpensive.

The receiver stage following the single-mode fiber is the most important and complex stage of a coherent optical communication system (Fig. 9.7). The optical mixer is at the heart of the receiver. The receiver is a four-port device where the two input optical fields (incoming and local) are mixed and added at the two output ports. After the optical-to-electrical conversion by the photodetector, the electrical currents are filtered by a band-pass filter in order to eliminate the constant d.c. term and to limit the additive noise. A portion of the photocurrent is fed back to lock the IF frequency to the desired value by means of automatic frequency control (AFC) loop. In homodyne and heterodyne systems, the AFC is required for the local oscillator in order to compensate for the slow frequency fluctuations of both the local oscillator and the transmitting laser. In a heterodyne receiver, a further electrical demodulation is performed before the signal output is fed to the decision-making circuitry.

The greatest advantage of a coherent (heterodyne/homodyne) optical communication system is the improvement of receiver sensitivity, that is, the reduction in the minimum required receiver signal level for achieving a prescribed bit error rate (BER). This improvement is attributed to two effects: (i) the improvement of the SNR, at the output end of the receiver preamplifier by the use of the heterodyne/homodyne scheme, and (ii) the improvement of the BER for a given SNR brought about by the use of a coherent modulation/demodulation scheme as compared to a non-coherent scheme. The heterodyne/homodyne schemes employed in optical communication systems are the only practical methods by which a nearly shot noise-limited detection has been realized. For coherent optical systems, in the case of binary digital transmission, the value of the quantum limit sensitivity, expressed in terms of required photons per bit (ph/bit), varies from 9 to 80 ph/bit to achieve a BER of 10^{-9}, depending on the modulation and the adopted detection techniques. If these values are compared with the quantum limited direct-detection sensitivity of 10 ph/bit, it can be argued that while IM-DD theoretical sensitivity is as good as, or better than, those of the coherent systems, actual sensitivities, affected by thermal noise, are typically a few tens of times larger than the theoretical value. The coherent detection experiments have demonstrated sensitivities within a factor two from the quantum limit. As a matter of fact, a coherent receiver sensitivity of 45 ph/bit has been achieved at a bit rate of 400 Mbit/s while at this bit-rate, a sensitivity of 450 ph/bit has been obtained for IM-DD systems.

Extensive research/development work on coherent optical communication systems has been carried out during the last decade because of the expectation for repeater-separation elongation and super wideband frequency-division multiplexing. Besides substantially improving the performance of the terrestrial optical fiber-communication links, the coherent communication systems are expected to revolutionize trans-oceanic links (less transmitter power requirement). Another major advantage of the coherent systems is the use of electrical filters rather than optical filters permitting densely-spaced FDM which require ultra-low line-width and frequency tunability.

9.6 Satellite optical communication systems

Well-established optical communication links offer many advantages vis-a-vis the microwave links already in use in the present satellite communication system. The main dampening effect on satellite optical links is the atmosphere, which seriously attenuates and scatters the optical beams passing through it. However, an optical free-space communication system for inter-satellite links and deep-space missions, where the atmosphere is nonexistent, has significant advantages in comparison to a microwave system. Further, high data rate inter-satellite optical communication links could take advantage of the new unoccupied optical frequency regions of the electromagnetic spectrum with much smaller antenna sizes. A possible scenario of inter-satellite links (ISL) is shown in Fig. 9.8. In such space-optical communication links, background (BG) radiation from natural sources reduces the SNR when they are in the field of view of the receiver, so that both data acquisition and tracking in ISLs suffer. Recent developments in coherent optical detection techniques will help solve some of the problems of high sensitive detection of optical signals in ISLs for improved data acquisition and tracking. This will naturally reduce the demand on the laser transmitter for power requirement and thus, along with a smaller antenna, simplifies the space-optical communication links.

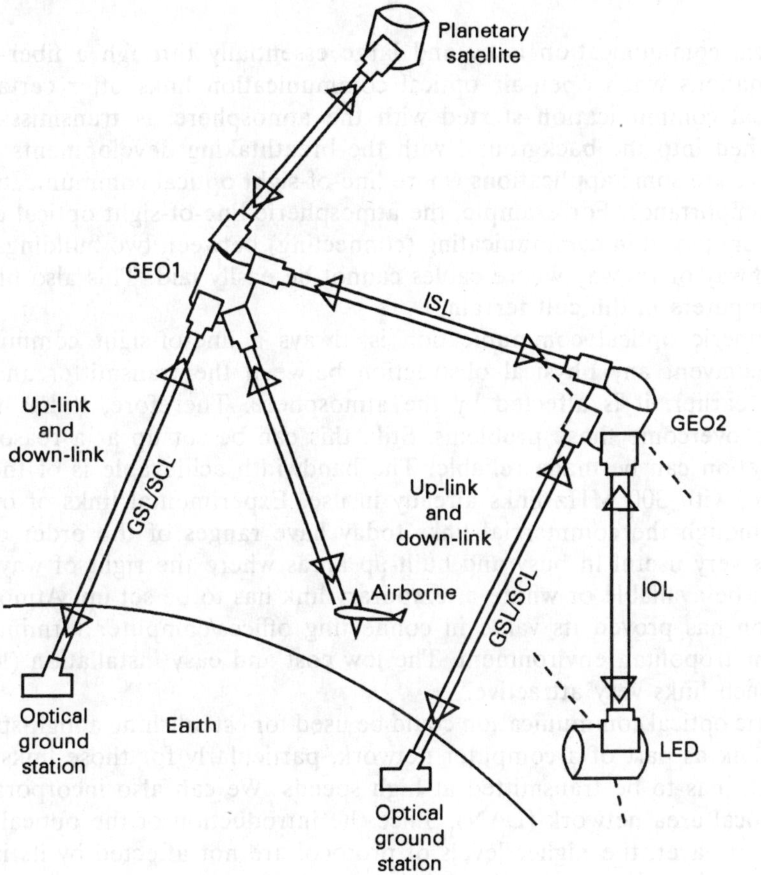

Fig. 9.8 *Configuration of inter-satellite laser communication links.*

Ground/aircraft-to-satellite and satellite-to-ground/aircraft optical communication links are receiving greater attention in recent years. Such links, however, are restricted by atmospheric turbulence, aircraft boundary layer effects, weather and ocean effects. In order to overcome the atmospheric effects, high-power laser systems have to be deployed for bridging the distances involved. Usually, a CO_2 laser is chosen for such links because it can deliver large power outputs efficiently at 10.6 μm, corresponding to the atmospheric window. Such a high-power laser system is fairly bulky and though it may not be a problem in installing it on the ground, it will demand considerable design skills to deploy it on board a satellite. The link also requires a highly sensitive optical receiver which must be coherent at 10.6 μm. The existing infrared detector technology at this wavelength is rather slow at this point of time but serious attempts are under way to make them faster. Besides, infrared detectors often have to be operated at low temperatures in order to overcome the intrinsic noise problems.

9.7 Terrestrial atmospheric optical communication lines

Though optical communication is by and large essentially through a fiber-optic medium, there are situations when open-air optical communication links offer certain advantages. Though optical communication started with the atmosphere as transmission medium, it has been pushed into the background with the breathtaking developments in fiber optics. However, there are some applications where line-of-sight optical communication links are of considerable importance. For example, the atmospheric line-of-sight optical communication links can be very useful in communicating (connecting) between two buildings on either side of a busy highway or runway where cables cannot be easily laid. This also finds application in linking computers in difficult terrain.

An atmospheric optical communication is always a line-of-sight communication since it cannot circumvent any physical obstruction between the transmitter and the receiver. As discussed earlier, it is affected by the atmosphere. Therefore, it has to be properly engineered to overcome these problems. Still, this can be set up at a reasonably low cost and the operation can be made reliable. The bandwidth achievable is of the order of 800 MHz or more, with 300 MHz links already in use. Experimental links of over 4 km have been set up though the commercial links today have ranges of the order of 1 to 1.5 km. Such a link is very useful in busy and built-up areas where the right of way tor laying the cable may not be available or where a temporary link has to be set up. Atmospheric optical communication has proved its value in connecting offices/computer terminals in high-rise buildings in metropolitan environment. The low cost and easy installation (less than half a day) makes such links very attractive.

Atmospheric optical communication could be used for establishing a high-speed computer-to-computer link as part of a computer network, particularly for those links where a large volume of data has to be transmitted at high speeds. We can also incorporate such a link as part of a local area network (LAN). Since the introduction of the optical link is only at the physical link layer, the higher levels of protocol are not affected by its introduction. A typical atmospheric link consists of a laser transmitter, transmitter optics/antenna, optical receiver (photodetector), receiver optics/antenna and the necessary modems. The most common method of data transmission using optical systems employs pulse-code modulation of a laser diode optical output. For short distances of the order of a few tens of metres, one might be able to get away with LEDs. The optical output from semiconductor devices is non-uniform and divergent, and requires suitable optics to collimate it. The receiver consists of a telescope type optics focusing the beam on a P-I-N/APD photodetector. A filter is placed in front of the detector to remove any extraneous light, and allows only the light at laser frequency to reach the detector. The receiver circuit also consists of a low-noise preamplifier.

The main sources of error are atmospheric aberrations like change in humidity and rainfall which could affect the data transmission considerably. Appropriate protocols at the hardware and software levels are required to correct it. At the hardware level, a variable clock is used for transmission and reception of data. Periodically, some standard frames are sent and if the bit error rate at any time is found to be too high, the modem automatically switches to a lower rate of transmission. At the lowest rate, the transmission is carried out one byte at a time and an acknowledgment is received after every byte. If there is a parity error, the transmission of the byte is repeated. If the weather improves, the bit error rate increases, and thus the transmission clock switches to a higher rate.

Other sources of error include material obstacles like flying birds cutting off the beam, and sunlight and noise in the receiver circuit. A multiple P-I-N diode receiver can be used to take care of the error caused by birds crossing the beam. The receiver telescope is so placed that direct sunlight does not reach the detector and that the intensity of scattered sunlight which will pass through the receiver filter is quite small. Also, sunlight is a slowly varying signal as compared to the transmitted signal and could easily be filtered out.

In summary, atmospheric optical links offer mobile data links at low cost. Their speed is high enough to be part of a LAN at 10 MHz. Even though the distances which can be achieved are less than 500 m using low cost components, we can very well expect the distance to grow to several kilometres in course of time, with improvements in lasers and detectors.

9.8 Underwater optical communication systems

The global communication network has two media, namely, satellite medium and submarine cable medium. The advances in the technology of both these media during the past two decades have been breathtaking. In certain aspects, it can be said that the improvements, new ideas and innovation in the submarine cable technology have surpassed those in satellite technology. Optical communication first gained entry into the submarine cable medium rather than the satellite medium. One of the reasons for this is that submarine cable technology has evolved over the last hundred years, whereas satellite technology is a relatively new phenomenon.

9.8.1 Fiber-optic trans-oceanic cable systems

The advances in digital technology and digital transmission based on PCM techniques has provided a base for introducing fiber optics as a transmission medium in underwater trans-oceanic cable systems. The main advantages of underwater fiber optical cable systems over conventional cables are:

1. High reliability due to extended repeater spacing.
2. Applicability of large capacity of submarine transmission system.
3. Easy handling characteristics like smaller cable diameter and light weight for installation and repairs.
4. Wide-band capability to handle large number of telephone, TV and data channels simultaneously.

Undersea optical fiber-based cable systems are generally used for two types of links:

(i) 565 Mbps channel capacity with a typical repeater spacing of 80–100 km.
(ii) 140 Mbps channel capacity with a typical repeater spacing of 120–150 km.

Since submarine cable systems generally span much greater distances than land-based systems, it is particularly important to achieve much larger repeater spacings. Besides, it is difficult and expensive to install and maintain a submerged repeater system. To meet these requirements, pure silica core low-loss dispersion-shifted optical fibers are used in conjunction with 1.55 μm buried heterostructure sources. In submarine cable systems, optical fibers are needed which combine low losses with low chromatic dispersion and have sufficient mechanical strength for use in demanding environment. Armoured fiber-optic cables are specially manufactured for deep-sea applications by providing additional protection depending on the conditions and nature of the sea bed. To take full advantage of the very low losses offered by single-mode fibers, it is necessary to minimize the ageing margin, i.e., the margin allowed for small changes in attenuation during the typical 25-year lifetime of a submarine cable system. This can be achieved only if OH (water vapour) diffusion into the fiber is minimized as the pressure of OH ions in the fiber drastically affects its attenuation.

There are several underwater fiber-optic cables already installed in several parts of the world which are working satisfactorily. The pace at which the trans-oceanic cable systems are being fibrized is increasing rapidly. In recent decades, there has been a great development in the field of trans-oceanic fiber-optic links. Due to their advantages like high bandwidth, large repeaterless spacing, etc., they score over their conventional counterparts. The first trans-oceanic fiber-optic link TAT-8 was installed in 1988 connecting US, France and Britain. Currently, eight trans-Atlantic fiber-optic cable systems like Amencas-I, Columbus-II, Carac, Bahamas, etc. connecting Europe and North America are either in operation or under construction. The Japanese have laid extensive underwater fiber-optic cables to connect their various islands, besides other forms of communication with neighbouring nations.

9.8.2 Optical communication with submerged bodies

Communication from the ground to submarines under water has always been a problem. This is because water is a good absorber of electromagnetic waves throughout the EM spectrum. The exceptions are VLF waves and blue green optical waves. Even with VLF waves, depth of penetration in water is not more than a few tens of metres. The submarine, therefore, has to come to the surface and risk being detected, to communicate with the ground, or should tow a buoy which is floating. This severely restricts the speed of the submarine and its manoeuvrability.

The blue green optical signal, on the other hand, has an e depth of 22 m in deep ocean waters. This implies that a signal with 1 W of peak power at the ocean surface can travel 300 m inside water before it falls below 1 μW, which is easily detectable. This is sufficient to conceive of a potential blue green optical link between ground and submarine.

As discussed, the maximum distance which even a blue green optical signal could travel in water is about 300 m. Thus, one cannot conceive of a meaningful direct line of sight communication between ground and submarine. The only meaningful link could be from ground to a mirror satellite, back to the ocean surface and then a penetration of few hundred metres of ocean water to reach the submarine as shown in Fig. 9.9(a). The problem of communication, is therefore,

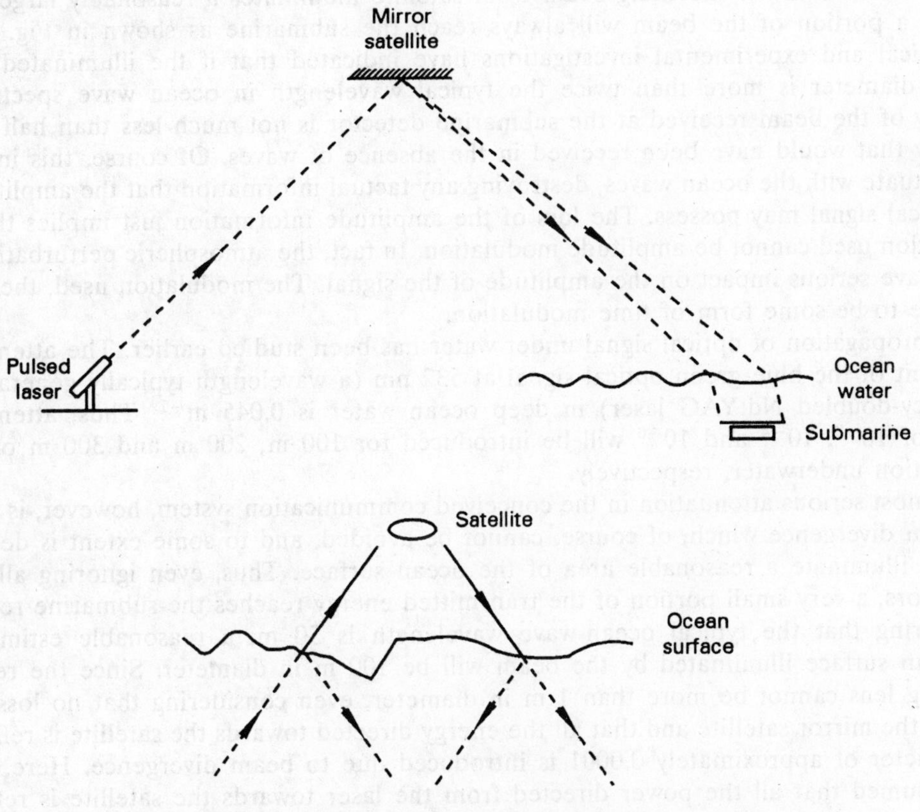

Fig. 9.9 (a) Ground to submarine optical links (b) Optical refraction at ocean surface.

(i) atmospheric communication from ground to satellite and from satellite to ocean surface.
(ii) penetration of air–water interface at ocean surface in presence of ocean waves, and
(iii) optical propagation in water from ocean surface to submarine.

Atmospheric optical communication has been reasonably studied for well over a decade. The attenuation of optical signal both due to absorption and due to scattering is primarily confined to a few hundred metres vertically up in the atmosphere. The contribution to the loss from higher atmosphere is nearly negligible. An attenuation factor of 0.36 is introduced due to the propagation of optical signal from ground to satellite and from satellite to ocean surface.

The signal penetration of ocean surface is considerably complicated by the presence of totally random ocean waves. Even ignoring the air bubbles, the ocean waves will cause the incoming beam to dance all over and refract into water at all angles. This dancing could be so severe that it would be a matter of luck if the optical signal is directed towards the detector placed in the submarine. The situation, however, is not as hopeless as it may appear at first sight. The ocean waves may be random but its wave spectrum still follows certain statistics. If the incoming beam from satellite illuminates a reasonably large ocean surface, a portion of the beam will always reach the submarine as shown in Fig. 9.9(b). Theoretical and experimental investigations have indicated that if the illuminated ocean surface diameter is more than twice the typical wavelength in ocean wave spectra, the intensity of the beam received at the submarine detector is not much less than half of the intensity that would have been received in the absence of waves. Of course, this intensity will fluctuate with the ocean waves, destroying any factual information that the amplitude of the optical signal may possess. The loss of the amplitude information just implies that the modulation used cannot be amplitude modulation. In fact, the atmospheric perturbation too would have serious impact on the amplitude of the signal. The modulation used, therefore, will have to be some form of time modulation.

The propagation of optical signal under water has been studied earlier. The attenuation coefficient of the blue green optical signal at 532 nm (a wavelength typically generated by frequency-doubled Nd:YAG laser) in deep ocean water is 0.045 m^{-1}. Thus, attenuation factors of 10^{-2}, 10^{-4} and 10^{-6} will be introduced for 100 m, 200 m and 300 m of wave propagation underwater, respectively.

The most serious attenuation in the conceived communication system, however, is due to the beam divergence which, of course, cannot be avoided, and to some extent is desirable so as to illuminate a reasonable area of the ocean surface. Thus, even ignoring all other loss factors, a very small portion of the transmitted energy reaches the submarine receiver. Considering that the typical ocean-wave wavelength is 50 m, a reasonable estimate of the ocean surface illuminated by the beam will be 100 m in diameter. Since the receiver collecting lens cannot be more than 1 m in diameter, even considering that no loss takes place at the mirror satellite and that all the energy directed towards the satellite is reflected, a loss factor of approximately 0.0001 is introduced due to beam divergence. Here, it has been assumed that all the power directed from the laser towards the satellite is reflected back and illuminates about 100 m of ocean surface. Assuming a further loss factor of 0.1 due to receiver filter and optics, the total attenuation factor introduced in the optical beam for submarines 100 m, 200 m and 300 m below water is tabulated in Table 9.1. If the receiver at the submarine could detect 1 μW of power, the laser power transmitted must be of the order of several kilowatts for the link to be established. This is only possible if a pulsed laser is used.

The power output of the available frequency-doubled Nd:YAG pulsed laser at 532 nm (blue green region) is of the order of 10^6 W or more. Such a laser could, potentially, be used for ground-to-submarine communication. But the pulse width of the optical signal is barely 10 ns and, therefore, not more than 50 pulses can be transmitted through the laser in a second. As already discussed earlier, the amplitude information of the optical signal is not preserved in the communication system and some form of pulse-time modulation is required to be used. A reasonable choice is pulse-interval modulation, where interval between the two pulses carries the information. In fact, using an Nd:YAG laser whose pulse-repetition rate can be varied from 20 to 50 pps, and dividing the interval between pulse in slots of 1 μs each, it is possible to transmit 14 to 16 bits of information with each pulse giving approximately 400 bits per second transmission rate. This is small, but could be increased with higher power laser designed with higher pulse-repetition rate.

In summary, communication between ground and a submerged submarine is quite difficult because of strong absorption of EM waves spectra by ocean water. The window of minimum attenuation is the blue green optical region. Even then the attenuation is high enough to restrict communication to less than a few hundred metres only. The only possibility is a communication system from ground to submarine *via* a satellite. Such communication is feasible using a high-power pulsed laser like Nd:YAG laser. The communication, however, is limited to submarines nearly 200 m below the ocean surface. Also the bit rate of communication possible as of now is so small that even a single voice channel cannot be transmitted. These restrictions may, however, ease with the development of new lasers.

9.9 Current developments

There are many current developments which augur well for the future of optical communications. We will only discuss some of the more promising developments.

9.9.1 Integrated optics

The development of integrated optical circuits is analogous to that of the development of VLSI technology in electronics. Integrated optics (IO) is the technology of planning numerous microscopic optical and electro-optical elements on a single plane of a common substrate. These devices typically are lasers, light waveguides, modulators, couplers, photodetectors and transistors. Basic to all of them is a thin transparent film wave guide for transmitting light energy. In fact, the name IO now covers all exploration of guided wave techniques used to construct new or improved optical/electronic devices. The birth of IO may be traced back to 1965 when it was first suggested that a microfabrication technology could be developed for single-mode optical devices with semiconductor and dielectric materials in a similar manner to that which had taken place with electronic circuits with all the inherent advantages of miniaturization. Fabrication of miniature thin film optical components such as couplers and lasers has been easier than the manufacture of complete integrated opto-electronic circuits on a single substrate.

Hybrid circuits have been built with a combination of discrete components and integrated circuits. Now the trend is towards building a monolithic integrated opto-electronic circuit on a chip. Opto-electronic circuits incorporating both optical sources and detectors along with single-mode optical waveguides have been successfully realized over the last few years. Just as integrated circuits have re-built and shaped the present state of electronics, IO is expected to have an equally major impact on the future of optical communication. The major advantages of IO circuits are: (i) lower driver-power requirements, (ii) compact (micro-miniaturization), (iii) inexpensive, (iv) faster, (v) very large bandwidth (high data-handling capacity), (vi) immunity to EMI, (vii) rugged, (viii) stable and (ix) efficient.

9.9.2 Soliton propagation in optical fibers

Soliton propagation is a relatively new field in nonlinear fiber optics and has attracted considerable attention due to its fundamental nature and commercial application. It was in 1973 that the possibility of soliton propagation in optical fibers was first mooted, and the experimental verification came in 1980. Since then, the field has grown rapidly and practical optical communication systems based on soliton propagation are being installed.

In soliton generation, the two factors which are most important are the material dispersion and the fiber nonlinearities. We have already seen that the pulse dispersion in optical fibers is the major contributory factor in limiting the bandwidth of an optical communication system. In single-mode fibers, material dispersion, which is sometimes referred to as chromatic dispersion or group velocity dispersion, is the main reason for limiting the bandwidth of the fiber-optic communication system. When the optical pulse intensity propagating through the fiber is strong enough, it can generate nonlinear effects in the fiber which can effectively cancel the material dispersion in certain wavelength regions. Thus, 1.55 μm (third optical window of the fiber) is a potential region for soliton generation and propagation. For a particular initial pulse shape (intensity distribution) provided by the optical transmitter (LED or LD), one gets the ideal case of soliton generation. Ideally, the soliton pulse once generated will propagate through the fiber with zero dispersion. Under certain conditions, even if the pulse changes shape in a complex manner, periodically at the end of a soliton period, it regains its initial shape. Another interesting property of solitons is the durability which means that once generated and launched into the fiber, they retain their shape over long distances.

Optical communication based on soliton transmission systems using erbium-doped fiber amplifier (EDFA) repeaterless links have been successfully developed to bridge trans-oceanic distances with bit rates up to 20 Gbps. However, soliton-based optical communication systems suffer from a number of constraints:

(i) The initial pulses must have a transform-limited squared hyperbolic secant (sech^2) form. A non-ideal pulse is able to reshape itself into a soliton pulse but it generates a small dispersive wave.
(ii) The required peak power for excitation of a fundamental soliton pulse in the fiber is given by

$$P_{\text{sol}} = \frac{0.776\lambda^2 A_{\text{eff}} D}{\pi^2 c \eta_2 t^2}$$

where A_{eff} = effective core area of the fiber in μm^2; D = chromatic dispersion in ps/m/nm; τ = width of soliton in ps; η_2 = nonlinear constant of fiber in m^2/W; λ = wavelength in m; c = speed of light in m/s.

(iii) In a lumped-amplifier (repeater) optical communication system, stable transmission can be achieved when the repeater span Z_a is sufficiently smaller than the soliton period Z_0 given by

$$Z_0 = 0.322\pi^2 c\tau^2/(\lambda^2 D) \tag{9.2}$$

(iv) The Gordon–Haus effect is the major limitation of soliton transmission. It is well known as a timing jitter resulting from random variation of the soliton centre frequency caused by spontaneous emission noise of optical amplifiers, fiber nonlinearity and the non-zero fiber dispersion.

(v) In soliton systems, the pulses interact with each other through the nonlinear index due to overlap of the tails of the pulses or through the radiated dispersive wave. This interaction also causes the jitter.

9.9.3 New materials and devices

One of the recent advances in optical fibers is the development of single-mode optical fibers which exhibit very low loss in the mid-infrared region above 2 μm since the Rayleigh scattering drops off very rapidly with increasing wavelength ($\propto 1/\lambda^4$). It is obvious that one has to operate at longer wavelengths in order to achieve low transmission losses in the fiber. However, in the case of silica fibers, the infrared absorption tail sets in quite early, resulting in a steep increase in attenuation beyond 1.55 μm. Thus, a new class of materials have to be developed which exhibit low losses in the mid-infrared region. This has led to the development of glasses based on fluoride compounds from which fibers have been drawn. The most promising amongst them are zirconium fluoride (ZrF_4) and beryllium fluoride (BeF_2) glasses whose attenuations at 2.5 μm are of the order of 0.02 dB/km and 0.005 dB/km, respectively. Since BeF_2 is highly toxic, most of the work on the development of low-loss fibers has been done using the material ZrF_4–BaF_2–LaF_3–AlF_3–NaF (ZBLAN). In ZBLAN fibers, the transmission window is at 2.55 μm when the attenuation coefficient is about 0.7 dB/km. The dispersion of ZBLAN fibers is highly dependent on the fiber structures, and dispersion factors of about 1 ps/nmkm have been achieved, which will enable high bit rate signals over long distances.

The move to higher wavelength is likely to see a new generation of lasers and detectors. The most promising semiconductor laser source is a double heterostructure single-mode laser diode based on InAsSbP in conjunction with InAs substrate. Besides, the requirement to handle high bit rates of information over long distances saw the development of special structures with higher power outputs. Though work in this direction is in a state of research and development, some practical devices are already available, and notable among them are the DFB (distributed feedback) and long external cavity (LEC) laser diodes.

In the case of a DFB laser diode, a periodic variation in refractive index in the laser heterostructure waveguide along the direction of wave propagation is provided so that the feedback of optical energy is obtained through reflection (see Fig. 9.10). Hence, the corrugated grating structure as shown in Fig. 9.10(a) determines the wavelength of the longitudinal mode emission rather than the Fabry–Perot gain curve of a normal LD as shown in Fig. 9.10(b). Thus, as observed in the figure, only one particular mode will lase while the other modes exhibiting higher losses are suppressed from oscillation. Extension of the laser cavity length by the introduction of external feedback can be achieved by using an external cavity into a wavelength dispersive element as part of the cavity. Such devices are often referred to as long external cavity (LEC) lasers. A wavelength dispersive element is required because the long resonator structure has very closely spaced longitudinal modes which necessitate additional wavelength selectivity. A diffraction grating is often used as an external mirror on the side of the cavity in order to filter out the lasing emission from wide-band gain spectrum of the laser diode, thus giving a narrow line-width at a desired wavelength.

For long wavelength detection, InGaAs photodiodes are generally used but lead sulphide (PbS) detectors are also showing promise. The optical FET (OPFET) is a hybrid opto-electric device based on a microwave GaAs FET, and has a response time of the order of 50 ps. Calculations show that FET geometry is nearly optimum for a high-speed photoconductor and has field response times lower than P-I-N or APD photodiodes. This detector is easy to integrate with optical circuits, requires low voltage, and is also reliable.

The InGaAsP photodiode which demultiplexes signals transmitted at two wavelengths, is used for WDM (wavelength division multiplexing)-based system. The wavelength discrimination at different wavelengths is achieved by the InGaAsP/InP multilayer structure that incorporates two InGaAsP layers of different bandgaps.

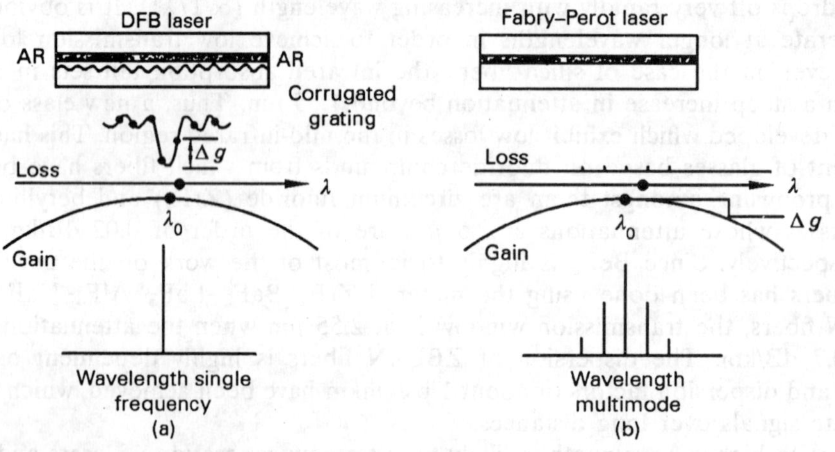

Fig. 9.10 The single-frequency operation of (a) DFB laser in comparison with (b) Fabry–Perot laser.

9.10 Future trends

The introduction of optical communication systems in global communication networks has brought about a quiet revolution in the quality (improved SNR) and quantity (channel capacity) of communications. Every new technology will be driven ultimately by the market requirement to successfully compete with the existing ones. Thus, the economics of implementing an optical communication system is getting better all the time with reduced component cost, due to mass production techniques. Further, the enhanced performance levels of the system both in terms of path length and bit rates makes it very competitive in terms of cost effectiveness. Thus, optical communication systems are either replacing or augmenting the existing communication channels all over the world at a hectic pace. Also the basic cost of the optical fiber can be very low compared to co-axial metallic cables because the raw material required for fibers is sand (SiO_2) which is abundantly available all over the world. This makes it also unattractive for pilfering.

Figure 9.11 shows how the repeaterless path length in a fiber-optic communication system falls off with the increasing bit rate. This implies a trade-off between maximum repeaterless distance and bit rate. In terms of technology development, five generations of optical communication are recognized as shown in Fig. 9.12. This figure shows the progressively increasing capabilities of each generation in terms of repeater spacing and data rates.

Bit-rate vs. distance for constant attenuation/km

(i) For an attenuation limit of 5 dB/km; the maximum distance varies from 15 km at 1 Mbps to 7 km at 1000 Mbps.
(ii) For an attenuation limit of 1 dB/km, the maximum distance varies from about 70 km at 1 Mbps to 30 km at 1000 Mbps.
(iii) The distance varies from 300 km for 1 Mbps to about 100 km for 1000 Mbps at an attenuation limit of 0.2 dB/km.

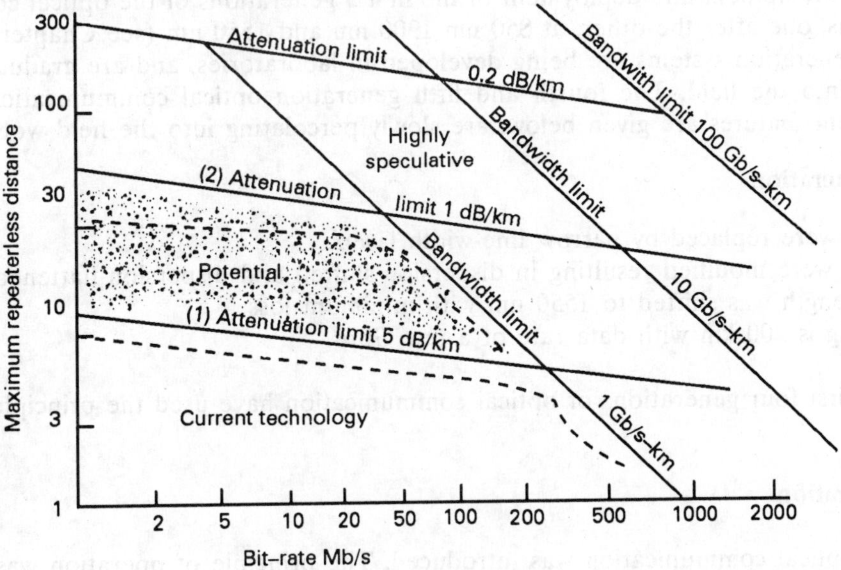

Fig. 9.11 *Bit rate vs distance for constant attenuation/km.*

The variation of distance with bit rate for various bandwidth limits can also be seen in a similar way.

(i) A bandwidth limitation of 1 Gbps corresponds to various optical properties of the fiber and depends on the optical wavelength and line-width of laser.
(ii) A bandwidth limitation of 10 Gbps corresponds to exploitation of laboratory results.
(iii) A bandwidth limitation of 100 Gbps corresponds to advances in technology in future.

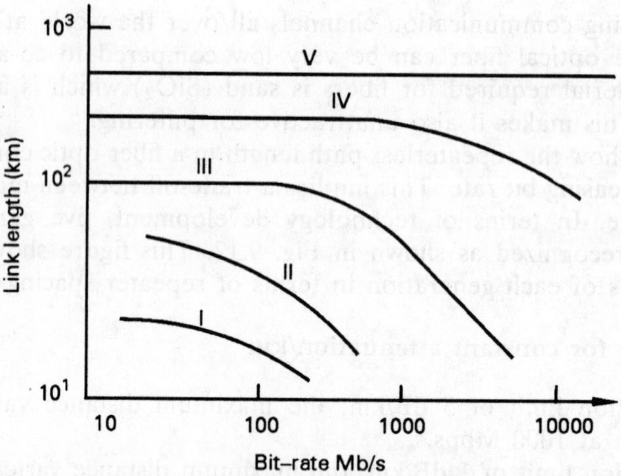

Fig. 9.12 The progressively increasing capabilities of each generation in terms of repeater spacing and data rates.

After the development and deployment of the first 3 generations of the optical communication systems, one after the other, at 850 nm 1300 nm and 1550 nm (see Chapter 1), fourth and fifth generation systems are being developed in laboratories, and are gradually finding their way into the field. The fourth and fifth generation optical communication systems, whose salient features are given below, are slowly percolating into the field worldwide.

Fourth generation

(i) LEDs were replaced by narrow line-width lasers.
(ii) Fibers were modified resulting in dispersion-shifted and dispersion-flattened fibers.
(iii) Wavelength was shifted to 1550 nm with minimum loss.
(iv) Spacing is 200 km with data rate of a few Gbps.

All the first four generations of optical communication have used the principle of direct detection.

Fifth generation

Coherent optical communication was introduced. The principle of operation was changed. Heterodyning technique was employed by mixing the received signal (before photodetection) with a very strong local oscillation signal.

The sensitivity of the receiver was substantially improved, and hence very large distances were achieved using very narrow line-width lasers.

The laboratory results for coherent communication have reported a distance of 200 km at 4 Gbps way back in 1988–89.

With the currently used silica fibers, it is difficult to achieve repeater spans beyond 500 km. Other new technologies are yet to be exploited.

Bibliography

In selecting the references for this book, the main consideration has been the fact that this book is mainly intended as a textbook on this subject at the undergraduate level of engineering. As such, references to research journals which are not normally available in such colleges have been avoided. Besides, while referring to books on this subject, care has been taken to select those which are normally stocked in engineering college libraries. Alternatively, they are easily available with book vendors at a relatively low cost. Many of these books, though originally published abroad, are now available in India as low-cost editions.

1. Sibley, M.J.N. (1995), Optical Communications: Components and Systems, 2nd Edition, Macmillan, UK.
2. Pratt, W.K. (1969), Laser Communication Systems, John Wiley & Sons, Inc., USA.
3. Amuon Yariv (1989), Quantum Electronics, 3rd Edition, John Wiley & Sons, Inc., USA.
4. Gowar, J. (1984), Optical Communication Systems, Prentice Hall International, UK. Third Indian Reprint published by Prentice-Hall of India Private Limited in 1992.
5. Palais, J.C. (1988), Fibre Optic Communications, 2nd Edition, Prentice-Hall, Inc., USA.
6. Steele, E.L. (1968), Optical Lasers in Electronics, John Wiley & Sons, Inc.
7. Cherin, A.H. (1983), An Introduction to Optical Fibers, McGraw-Hill International Book Company (International Student Edition).
8. Mukunda Rao, M. and Ashok Jhunjhunwala (Eds.), Optical Communication: Proceedings of the International Workshop held at IIT Chennai during March 7-12, 1985.
9. Kao, C.K. (1982), Optical Fibre Systems: Technology, Design and Applications, International Edition, Published by McGraw-Hill, Inc.
10. Franz, J. and Jain, V.K. (1996), Optical Communication Systems, Narosa Publishing House, India.
11. Senior, J.M. (1994), Optical Fibre Communications: Principles and Practice, 2nd Edition, Prentice-Hall of India Pvt. Ltd., India.
12. Agrawal, G.P. (1992), Fibre-optic Communication Systems, John Wiley & Sons, Inc., USA.
13. Williams, E.W. and Hall, R. (1978), Luminescence and the Light Emitting Diode, Pergamon Press, UK.
14. Bishnu P. Pal (Ed.) (1992), Fundamentals of Fibre Optics in Telecommunication and Sensor Systems, Wiley Eastern Limited, India.
15. Allard, F.C. (1990), Fibre Optics Handbook for Engineers and Scientists, McGraw-Hill Book Co., International Edition.

16. Verdeyen, J.T. (1993), Laser Electronics, 2nd Edition, Prentice-Hall of India Pvt. Ltd., India.
17. Ajoy Ghotak and K. Thyagarajan (1993), Optical Electronics, Foundation Books of India by arrangement with Cambridge University Press.
18. Das, P. (1991), Laser and Optical Engineering, Narosa Publishing House, India, Springer International Student Edition 1991.
19. Yu. Yakushenkey (1983), Electro-optical Drives: Theory and Design, Mir Publishers, Moscow.
20. Sharma, A.B., Halme, S.J., and Butusov, M.M. (1981), Optical Fibre Systems and Their Components, Springer-Verlag, Germany.
21 Rampal, V.V. (1992), Photonics: Elements and Devices, A.H. Wheeler & Co. Ltd., India.
22. Thyagarajan, K. and Ghatak, A.K. (1984), Lasers: Theory and Applications, Macmillan India Limited.
23. Ghatak, A.K., Culshaw, B.,Nagarajan, V., and Khurana, B.D. (Eds.) (1995), Trends in Fibre Optics and Optical Communications, Viva Books Pvt. Ltd., India.
24. Pallab Bhattachary (1995), Semiconductor Optoelectronic Devices, Prentice Hall of India Pvt. Ltd., India.
25. Yu. Lyubavsky and V. Ovchinnikov (1975), Solid State Laser Technology, Mir Publishers, Moscow.
26. Tarasov, L.V. (1983), Laser Physics, Mir Publishers, Moscow.
27. Mukunda Rao, M. and Sirohi, R.S. (Eds.) (1984), Laser Techonolgy and Applications, COSTED, India.
28. Killen, H.B. (1991), Fibre Optic Communications, Prentice Hall.
29. Haus, H.A. (1984), Waves and Fields in Optoelectronics, Prentice Hall.
30. Agarwal, D.C. (1988), Fibre Optic Communication, A.H. Wheeler & Co. Pvt. Ltd., India.

Physical constants

Velocity of light in vacuum	c	$299,792,456\ \text{ms}^{-1} \approx 3 \times 10^{8}\ \text{m s}^{-1}$
Planck's constant	h	$6.6262 \times 10^{-34}\ \text{J s} = 4.1357 \times 10^{-21}\ \text{MeV s}$
Planck's constant	$\hbar$	$1.0546 \times 10^{-34}\ \text{J s} = 6.5822 \times 10^{-22}\ \text{MeV s}$
Boltzmann's constant	k	$1.3806 \times 10^{-23}\ \text{J K}^{-1} = 8.617 \times 10^{-11}\ \text{MeV K}^{-1}$ $= 3.300 \times 10^{-24}\ \text{cal K}^{-1}$
Gravitational constant	G	$6.67 \times 10^{-11}\ \text{Nm}^{2}\ \text{kg}^{-2}$
Permittivity of free space	ϵ_0	8.8542×10^{-12} farad m^{-1} or $\text{C}^{2}\text{N}^{-1}\ \text{m}^{-2}$
Permeability of free space	μ_0	$4\pi \times 10^{-7} = 12.5664 \times 10^{-7}$ henry m^{-1}
Elementary charge	e	1.6022×10^{-19} C
Bohr magneton	μ_B	$9.2741 \times 10^{-24}\ \text{J}(\text{Wb/m}^{2})^{-1}$
Avogadro's number	A	6.02217×10^{23} (gram-) mole^{-1}
Absolute zero temperature	0 K	$-273.16\,^{\circ}$C
Standard acceleration by gravity	g	$9.80665\ \text{m s}^{-2}$

Index

absorption coefficient, 44, 59
absorption loss due to impurities, 85
acceptance angle, 69
acceptance cone, 69
amplifier noise, 11
angular spreading, 62
atmospheric attenuation, 59
atmospheric propagation, 62
atmospheric turbulence, 61
attenuation, 62
avalanche photodetector (APD), 111
avalanche photodiode, 11

background radiation-limited operation, 104
beam collimator, 51
beam expander, 51
beam scintillation, 61
beam spreading, 61
Bessel function, 77
bit error rate (BER), 71

cladding refractive index, 70
coherence, 31
coherent detection, 12
coherent optics fiber transmission, 15
core refractive index, 70
critical angle, 69
CW mode, 35

D-star, 104
dark current, 104
detectivity, 104
diode
  quantum efficiency, 11
directionality, 31
distortion, 86
distributed feedback lasers, 50

electromagnetic absorption, 18
electromagnetic spectrum, 3
energy-level diagram, 29
external cavity semiconductor laser diode, 50

Fabry–Perot cavity, 4
fiber, 87
  attenuation characteristics, 87
fiber-optic communication link, 63
frequency response, 104

group delay, 71
guided wave propagation, 7

hole–electron pair, 43

image dancing, 61
impulse response, 104
infrared absorption tail, 85
injection laser, 4
integrated optical circuits, 15
integrated optics, 15
internal detector noise-limited operation, 104
intrinsic material absorption loss, 85

laser, 2
laser diode, 4, 6, 40
  requirements, 4
light emitting diode, 6
maser, 17
material dispersion, 8
microbending loss, 85
monolithic integrated optical circuits, 14

noise equivalent power, 104
numerical aperture, 70

optical communication
  LEDs, 6
  sources, 4
optical fiber
  attenuation, 9, 87
  characteristics, 8
optical heterodyne detection, 13
optical resonator, 27
optical resonator cavity, 29
optical transmission systems, 13
optical transmitting antenna, 51
optical waveguides, 8, 9, 13, 14
optoelectronic integrated circuit (OEIC), 16
optoelectronic transducer, 13
P-I-N photodiode, 11
photodetectors, 103
photodiode, 11
photomultiplier tube (PMT), 111
photophone system, 1
population inversion, 24, 29
pulse delay, 86
pulsed code modulation, 71

Rayleigh scattering loss, 85
receiver sensitivity, 11
regenerators, 14
relative refractive index difference, 69
repeaters, 14
round-trip gain, 25
ruby laser, 35

scattering coefficient, 59
semiconductor laser diode, 35, 50
signal attenuation, 64
silica glass, 9
single-mode fiber, 86
spatial coherence degradation, 61
spatial spreading, 62
spectral purity, 31
spectral response, 104
spontaneous emission, 18
step-index fiber, 71
stimulated emission, 18
sustained oscillations, 25

terrestrial optical communication, 6
  links, 6
thermal excitation, 37
threshold gain, 25
transmission capacity, 14
transmitting antenna, 51

ultraviolet absorption tail, 85

waveguide scattering loss, 85